Western Blotting Guru

Western Blotting Guru

Ayaz Najafov
Harvard Medical School, Department of Cell Biology, Boston, MA,
United States

Gerta Hoxhaj
Harvard T.H. Chan School of Public Health, Department of Genetics and
Complex Diseases, Boston, MA, United States

ACADEMIC PRESS

An imprint of Elsevier

Academic Press is an imprint of Elsevier
125 London Wall, London EC2Y 5AS, United Kingdom
525 B Street, Suite 1800, San Diego, CA 92101-4495, United States
50 Hampshire Street, 5th Floor, Cambridge, MA 02139, United States
The Boulevard, Langford Lane, Kidlington, Oxford OX5 1GB, United Kingdom

British Library Cataloguing-in-Publication Data
A catalogue record for this book is available from the British Library

Library of Congress Cataloging-in-Publication Data
A catalog record for this book is available from the Library of Congress

ISBN: 978-0-12-813537-2

For Information on all Academic Press publications
visit our website at https://www.elsevier.com/books-and-journals

  Working together
to grow libraries in
ELSEVIER Book Aid International developing countries

www.elsevier.com • www.bookaid.org

Publisher: Mica Haley
Acquisitions Editor: Linda Versteeg-Buschman
Editorial Project Manager: Timothy Bennett
Production Project Manager: Poulouse Joseph
Cover Designer: MPS

Chapter header images were engraved by Elizabeth Blackwell ca. 1737. Available from the NLM Digital
Collections at http://resource.nlm.nih.gov/101456746, NLM Image ID C03093.

Typeset by MPS Limited, Chennai, India

CONTENTS

Preface vii
Note to the Reader ix

Chapter 1 Introduction 1
1.1 What Is Western Blotting? 1
1.2 A Bit of History 2

Chapter 2 Procedure 5
2.1 Sample Preparation 6
2.2 Gel Electrophoresis 6
2.3 Protein Transfer 10
2.4 Blocking the Membrane 13
2.5 Primary Antibodies 14
2.6 Secondary Antibodies 17
2.7 Blot Washes 18
2.8 Developing Western Blots 18

Chapter 3 Good Practices 21

Chapter 4 Optimization and Troubleshooting 29
4.1 Optimization Rules 29
4.2 General Optimization Strategies 30
4.3 Troubleshooting Specific Problems 33

Chapter 5 Tips and Tricks 41

Chapter 6 Special Cases 51
6.1 Quantitative Western Blotting 51
6.2 Overlay Assays 52
6.3 Phospho-Specific Antibodies 52
6.4 Phos-Tag 53
6.5 Nonreducing PAGE 54
6.6 Dot Blots 55

Chapter 7 Data Analysis, Storage, Retrieval **57**

Appendix A: Buffers and Solutions...**63**
Appendix B: SDS−PAGE Gel Tables..**67**
Appendix C: SDS−PAGE Protocol **69**
Appendix D: Wet Transfer and Immunoblotting Protocol...................**71**
Appendix E: Home-Made Enhanced ChemiLuminescence (ECL)
Detection..**73**
Appendix F: Stripping Protocols...**75**
Appendix G: Coomassie Staining Protocol **77**
Appendix H: Lysis of Cells Using Native Conditions..........................**79**
Appendix I: Quick Denaturing Lysis Protocol.....................................**81**
Appendix J: Protein Tags..**83**
Appendix K: Covalent Crosslinking of Antibodies to Beads **87**
References **89**

PREFACE

This book was written with an aim of providing biologists with a handy reference and guide for all aspects of performing Western blotting and obtaining consistent, reliable, high-quality data. We believe that this book will provide tremendous assistance for setting up Western blotting systems and solving problems associated with challenging cases. The intended target audience is both beginners and experts.

We did not delve in scrupulous details of the gel electrophoresis and immunoblotting theory. Instead, this book is a *laboratory* guide, and the emphasis here is on the technical aspects of employing Western blotting as a tool in molecular biology laboratories. The issues were described in "bench terms," which makes the book more relevant to the end users of the technique.

General optimization and troubleshooting approaches and strategies, as well as a detailed guide for addressing specific Western blotting problems, tips and tricks developed and learned through years of experience with Western blotting, several special cases for Western blotting, and appendices with detailed protocols, as well as other useful information related to Western blotting, make this book a practical reference all molecular biology laboratories can benefit from.

The terms "Western blotting" and "immunoblotting" will be used interchangeably in the text. We will use the term "target protein" for the protein of interest to be analyzed by Western blotting. "Room temperature" will be routinely abbreviated as "RT" and "molecular weight" will be routinely abbreviated as "MW."

All incubations for Western blotting membranes are to be done preferably on a side-to-side shaker/rocker. Radial shakers/rockers are not the top choice, since they tend to result in an uneven washing of the membranes.

Phospho-specific antibodies will be given primary consideration in the text. However, most of the discussed concepts can be applied to antibodies specific for other posttranslation modifications of proteins, such as methylation and acetylation.

Always wear appropriate personal protective equipment!

Note: Acrylamide is an acute toxin and a health hazard.

Note: Ponceau S is an irritant. Use necessary personal protective equipment.

Note: Acetic acid is flammable and corrosive.

CHAPTER 1

Introduction

Only those who dare, drive the world forward.

—*Cadillac, Dare Greatly*

The superior man is satisfied and composed; the mean man is always full of distress.

—*Confucius*

1.1 WHAT IS WESTERN BLOTTING?

Western blotting (or immunoblotting) is a widely used method for protein detection, using antibody-based probes to obtain specific information about target proteins from complex samples. It is a routine method in molecular biology, biochemistry, and cell biology fields with a multitude of applications. This method can be used to obtain information about quantity, molecular weight, and post-translational modifications of proteins. Due to high affinities of antibody toward their epitopes and amplificatory nature of Western blotting, it is a very sensitive method and even picogram quantities of

Western Blotting Guru. DOI: http://dx.doi.org/10.1016/B978-0-12-813537-2.00001-1

target proteins can be detected. Depending on the type of reagents used, Western blotting can be quantitative or semiquantitative (see Chapter 6: Special Cases for quantitative Western blotting).

1.2 A BIT OF HISTORY

The ancestral versions of the modern Western blotting were first described by two laboratories in 1979:Harry Towbin and colleagues at the Friedrich Miescher Institute, Basel/Switzerland and George Stark and colleagues at Stanford University, California/USA. A more improved version of the method was developed and the name "Western blotting" was given in 1981 by W. Neal Burnette at the Fred Hutchinson Cancer Research Center Washington/USA, simply because of the location of the laboratory on the west coast of the USA, as a word play to the previously invented Southern blotting method for DNA detection, in 1975, and Northern blotting method, for RNA detection, in 1977. According to Google Scholar, Towbin's landmark paper has been cited more than 54,000 times in the last 38 years (Figures 1.1−1.3).

Figure 1.1 The inventors of the western blotting.Left to right: Harry Towbin, W. Neal Burnette, George Stark.

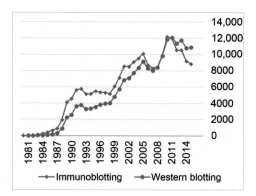

Figure 1.2 Number of articles found per year when indicated keywords were searched in PubMed (https://www. ncbi.nlm.nih.gov/pubmed/).

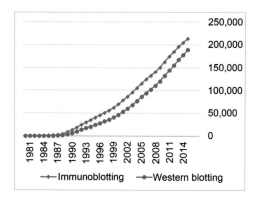

Figure 1.3 Number of articles accumulated over the years, when indicated keywords were searched in PubMed (https://www.ncbi.nlm.nih.gov/pubmed/).

CHAPTER 2

Procedure

The young do not know enough to be prudent, and therefore they attempt the impossible—and achieve it, generation after generation.

—Pearl S. Buck

In a moment of decision, the best thing you can do is the right thing to do. The worst thing you can do is nothing.

—Theodore Roosevelt

Inspiration usually comes during work, rather than before it.

—Madeleine L'Engle

There are five major steps in the Western blotting procedure (Fig. 2.1), and every step is critical for obtaining high-quality, reliable, and analyzable data.

Figure 2.1 Five major steps of western blotting.

Western Blotting Guru. DOI: http://dx.doi.org/10.1016/B978-0-12-813537-2.00002-3

2.1 SAMPLE PREPARATION

Cell or tissue lysates can be prepared in several different ways (see Appendices H and I), using either denaturing or native lysis buffers. Protein samples are best prepared using fresh lysis buffers in the presence of protease and phosphatase inhibitors to avoid the degradation of target proteins by proteases or dephosphorylation, if one is interested in probing for phosphorylated proteins. Protein samples should always be kept on ice to minimize degradation and dephosphorylation. For native lysate storage (e.g. to preserve protein activity or protein complex assembly), 0.27 M sucrose- or 10% glycerol-containing samples can be snap-frozen in liquid nitrogen and stored at $-80°C$. After clearing the lysates by centrifugation at $16,000 \times g$, 15 minutes, 4°C, protein concentration can be determined via Bradford protein assay or Bicinchoninic acid assay (BCA). BCA is less variable and less susceptible to detergents than Bradford, while Bradford assay has a simpler and faster procedure. After the protein concentrations are quantified, the samples can be prepared for loading into the gels by supplementing the lysates with $5 \times$ sodium dodecyl sulfate–polyacrylamide gel electrophoresis (SDS–PAGE) sample buffer to achieve a final concentration of $1 \times$ (see Appendix A). $1 \times$ SDS–PAGE sample buffer can be used to equalize all the protein and sample buffer concentrations across samples.

One important aspect of sample preparation is the final protein concentration of the samples. Typically, $10-50 \, \mu g$ of total lysate protein per well should be loaded. Thus, the sample concentration should be between 1 and 2.5 mg/mL, in order to be able to load $10-20 \, \mu L$ per well. Samples should be normalized to have the same concentration, so that the loading volume is consistent across the samples in a gel. If the loading volume changes between samples, it will affect the running of each sample. For example, a higher volume sample can interfere with the running of a sample that has low volume. After the samples are prepared, they should be denatured and spun down (see Appendix C). Proteins in sample buffer can be stored at $-20°C$.

2.2 GEL ELECTROPHORESIS

SDS–PAGE is a method that allows resolution of denatured proteins by their molecular weight (MW), using an electric field and a porous acrylamide-based matrix. To prepare the proteins for gel electrophoresis, the protein extracts are mixed with SDS, an anionic detergent that binds uniformly to proteins—approximately one SDS molecule for every two amino acid residues. Addition of SDS results in disruption of the tertiary structure of proteins into linear molecules and proteins become coated with a negative charge. To further unfold proteins, thiol (−SH) group containing compounds such as β-mercaptoethanol or dithiothreitol are also employed in order to reduce the disulfide bonds present in many proteins. Finally, samples are heated to 70−95°C to maximize the denaturation and enhance the uniformity of coating by SDS.

The mobility of each molecule is dependent on the net charge of the molecule, the voltage of the field and the resistance of the solution in which is immersed. Negatively charged, SDS-treated proteins will migrate through a polyacrylamide gel matrix toward the positive anode, with smaller proteins migrating faster than larger proteins.

Polyacrylamide is an ideal choice for the gel matrix because it is chemically inert, thermo-stable, transparent, and most importantly can produce a wide range of pore sizes, which allows for the separation and resolution of the proteins of interest. The polyacrylamide matrix is strong and can withstand high voltage, can be subjected to various staining and destaining procedures, as well as drying for autoradiography.

One of the most important aspects of the polyacrylamide gel is the pore size, which is determined by the amount of total acrylamide ($\%T$) and the amount of crosslinker bis-acrylamide ($\%C$) present in the gel.

A high $\%T$ gel has smaller pores and is used for resolution of small proteins. The most commonly used acrylamide:bis-acrylamide ratio for SDS−PAGE is 37.5:1 ($\%C = 2.6$), whereas the 19:1 gels ($C = 5\%$) are used for denaturing DNA/RNA gels and 29:1 gels ($C = 3.3\%$) are used for native DNA/RNA gels.

SDS−PAGE gels usually employ a two-part (discontinuous) gel system with a stacking gel on top of the resolving gel. Stacking gels have lower $\%T$ (4%−5%) and a pH of 6.8, minimizing the charge on the proteins, thereby allowing them to "stack" on top of each other, resulting in sharper bands. Resolving gels have higher $\%T$ (6%−15%) and a pH of 8.8. Some precast commercial gel systems use a different buffering system without stacking gels, since the gels have a gradually increasing %*T* from the combs to the bottom of the gel (4%−20%) giving a gradient nature to the gel.

2.2.1 Gel Selection and Preparation

Gel system type and gel percentage selection is an important decision for obtaining optimal results. The gel percentage choice depends on the size of the protein(s) of interest. The table below gives a range of percentages depending on the MW of the proteins (Table 2.1).

Either commercial precast or hand-cast gels can be chosen. Hand-cast gels, in addition to being a low cost product, can be readily customized based for a few parameters such as chemical formulation of gels, number of wells, and gel thickness. Hand-cast gels are poured between two glass plates (see Appendices B and C for detailed protocols) and a comb is placed on the top of the gel to create wells for loading the samples. The number and size of the wells, as well as the thickness of the gel determines the number of samples and volume that can be loaded into the gels.

Precast gels are offered commercially by many leading companies such as Bio-Rad (Mini-PROTEAN Precast Gels) and Thermo (The NuPAGE Precast Gel System) and they too come in different sizes, percentages and formulations. If the researcher is interested in probing for multiple proteins with major differences in their MW, gradient gels are a good choice. In other words, gradient gels (4%−12%, 4%−15%, 4%−20%) offer the benefit of detecting large and small proteins in the same sample, on the same blot.

The small gels usually have 10−15 well sizes. If the number of samples exceeds that, Bio-Rad offers the Criterion gel system with a variety of gel formulations on which 18 or 26 samples can be run simultaneously. One feature of The NuPAGE Precast Gel System is

Table 2.1 Recommended Gel Concentrations For Various Protein Size Ranges	
% gel (acrylamide:bis = 37.5:1)	Recommended Size Range
6	60−250 kDa
8	50−200 kDa
10	25−100 kDa
12	15−80 kDa
15	14−60 kDa

that the gel formulation does not contain SDS, but the denaturing conditions are simulated by the acidic pH of the buffer and the SDS in the running buffer.

2.2.2 Running the Gels

Before running the gels, the combs are carefully removed and the wells are either rinsed or aspirated to clean them from any potential semi-polymerized acrylamide debris. To improve the accuracy of the loading, special elongated gel loading tips can be used. The wells should definitely be underloaded, as accidental overflow from one well to another can easily alter the results. The standard way to run gels is to run them until the bromophenol blue dye front reaches the bottom of the gel. However, depending on the required resolution, the gels can be run longer. In such cases, attention should be paid to keep the running tanks cool, which can be easily achieved by half-submerging them into ice-water bath.

If when running a hand-cast gel, usually the gel is run at low voltage (70–80V) at the beginning so that the proteins enter the stacking gel with minimal distortion of the bands. As a rule of thumb, 10–15 V per cm of gel should be used for resolving gels, which is usually about 100 V for the commonly used "mini" gels. Gels should be ran slower when high-concentration or high-viscosity samples are ran, in order to avoid streaking and misshapened bands. However, due to novel tank designs, some commercial tanks allow running the gels at voltages as high as 200V from the beginning of the run (e.g. Mini-PROTEAN® Tetra cell from Bio-Rad).

2.2.3 Molecular Weight Markers

A protein MW marker is a mixture of highly-purified proteins which resolve as clearly identifiable ladder of bands usually between 10 and 250 kDa. Two types of MW markers are commercially available. **Prestained** MW markers give the advantage to monitor protein resolution in the gels in real time and determine transfer efficiency without the need to stain with Ponceau S, but are more expensive and sometimes do not have a very accurate or consistent MW estimation. **Unstained** MW markers are cheaper and give a more precise MW estimation, but cannot be visualized during the gel run and need to be stained by Ponceau S and marked by a pen/pencil on the membrane. Dual color prestained markers are usually the best choice, since it is easier to identify which band is represents which MW.

2.3 PROTEIN TRANSFER

2.3.1 Transfer Setup

Proteins need to be electrotransferred onto either a nitrocellulose or poly-vinylidene difluoride (PVDF) membrane, in order to increase the robust-ness of the protein-carrying matrix (i.e. membranes are more robust than gels), simplify the handling process and provide greater epitope accessi-bility to the antibodies (i.e. membranes are thinner than gels).

The first decision that needs to be made is the type of transfer: wet versus semi-dry. Although the semi-dry transfer is faster than the wet transfer and requires smaller volumes of transfer buffer, the latter offers better transfer efficiency (especially for large proteins), as well as higher and consistent quality of transfer. Thus, the rest of the chapter will be generally relevant to the wet transfer type, while specifics for semi-dry transfer protocol can vary between different transfer machines and should be therefore obtained from their manuals.

To transfer proteins, a typical sponge-paper-gel-membrane-paper-sponge transfer sandwich is assembled (Fig. 2.2). Attention should be paid after adding the membrane on the top of the gel to ensure that there are no air bubbles. The gel can be first immersed in transfer buffer, and after adding the membrane on top of the gel, the poten-tially trapped air bubbles should be rolled out using a roller or a sero-logical pipette. The sandwich should be closed and clamped with the transfer cassettes tightly and carefully, ensuring that components of the sandwich do not shift. The sandwich is then submerged into

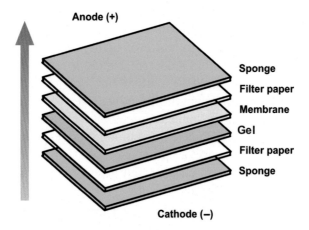

Figure 2.2 Diagram of a typical western blotting transfer sandwich setup.

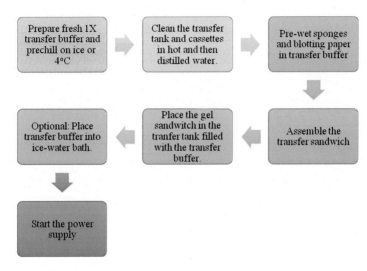

Figure 2.3 Steps of the protein transfer stage of western blotting.

transfer buffer. The proteins are negatively charged due to residual SDS left from the running on the SDS−PAGE gel and will travel from negatively charged cathode toward positively charged anode when an electric field is applied (Fig. 2.3).

2.3.2 Choosing the Right Membrane

Two common membrane types used for transferring proteins are nitrocellulose and PVDF, since they have a high binding capacity for proteins. One important characteristic of the membranes is the pore size, which is usually 0.1, 0.2, or 0.45 μm. Low pore size is recommended for low MW proteins (<20 kDa), whereas for most of the proteins 0.45 μm pore size can be used. When protein quantification is critical, smaller pore membranes should be chosen. The binding capacity for nitrocellulose and PVDF membranes is $80-100$ μg/cm^2 and $170-200$ μg/cm^2, respectively. **Therefore, for low-abundance proteins or low-affinity antibodies, PVDF can be chosen to increase sensitivity.** On the other hand, nitrocellulose membranes have lower overall backgrounds than PVDF membranes.

PVDF membranes are highly hydrophobic and need to be submerged in 100% methanol or ethanol prior to wetting in transfer buffer. PVDF membranes are a great choice for detecting hydrophobic proteins, such as lipidated (e.g., LC3) or transmembrane proteins (e.g., receptors). PVDF membranes are less fragile than nitrocellulose

Table 2.2 Comparison of Nitrocellulose and PVDF Membranes		
	Nitrocellulose Membranes	PVDF Membranes
Binding capacity	80–100 µg/cm^2	170–200 µg/cm^2
Nonspecific background	Lower	Higher
Autofluorescence background	Lower	Higher
Fragility	Fragile	Durable
Hydrophobicity	Lower	Higher
Ponceau S staining	Good	Low
Rounds of reprobing	Fewer	More
Choose for lipidated proteins	No	Yes
Choose for glycoproteins	Yes	Maybe

membranes and they can be easily stripped and reprobed without loss of sensitivity.

Most PVDF membranes cannot be effectively stained by Ponceau S and instead UV trans-illuminator can be used to visualize proteins after transfer. PVDV membranes usually have high autofluorescence and if using fluorophore-conjugated antibodies, special PVDF should be used.

Nitrocellulose membranes are easily stained by Ponceau S to visualize the transfer of the proteins and they should only be submerged in transfer buffer prior to use (i.e. no need for methanol treatment). Nitrocellulose is also a good choice for binding glycoproteins and for detection using fluorophore-conjugated antibodies (Table 2.2).

2.3.3 Blotting Paper

Blotting paper needs to be thoroughly pre-wet in the transfer buffer before setting up the transfer sandwich, in order to prevent air bubble formation during the transfer. Floating the blotting papers on the surface of the transfer buffer one-by-one ensures that air inside the paper is given enough time to escape the paper. Blotting paper is an important component of the transfer system because it helps the moving of the transfer buffer through the gels, to facilitate the movement of the proteins out of the gel and onto the membrane. Too thick blotting paper (or too many layers of blotting paper) can result in an over-squeezing of the gels and membranes and thus, distorted bands. Too thin/few blotting papers may result in incomplete/non-uniform transfer due to lack of full contact between the membranes and the gels.

2.3.4 Transfer Buffer

The common transfer buffer formulation is: 25 mM Tris–HCl pH 8.3, 192 mM glycine, 20% (v/v) methanol or ethanol. Although methanol is cheaper than ethanol, it is also more toxic. SDS is not included in the transfer buffer, as the SDS present in proteins is sufficient to move them to the membrane. SDS presence in the transfer buffer can cause a blowout, a condition in which the proteins do not bind to the membrane but instead pass through it. The presence of methanol or ethanol in the transfer buffer helps to increase the binding of proteins to the membranes, as they strip off SDS from proteins. In cases where the protein is known to be prone to precipitation or for high MW proteins ($>$150 kDa), SDS can be added (0.05%–0.1%). For efficient transfer of high MW proteins, CAPS-based buffer (10 mM CAPS, pH 11, 10% methanol) can also be used.

2.3.5 Transfer Power Settings

Depending on the protein MW, gel percentage, transfer apparatus and buffer system used, different power settings might be required. A typical setting for transfer is constant 100 V for 1 hr. However, for high MW, longer transfers, and especially for larger tanks, constant 0.4 A for 2 hours should be employed, concomitant with immersion of the tank into ice-water bath at least up to the middle of the tank height. Transfer can be set up with a timer (e.g., for 2 hours) and left overnight at RT to be continued with next day.

2.3.6 Visualization of Proteins After Transfer

To monitor the transfer quality and efficiency, proteins can be easily visualized by briefly (30–60 seconds) incubating the membranes in Ponceau S dye solution [0.1% (w/v) in 5% acetic acid]. The background is destained via several tandem rinses with distilled water. Ponceau S staining is also a good visual aid for accurate membrane cutting/trimming. The dye is completely removed by briefly washing the membranes in Tris-Buffered Saline Tween 20 (TBST). If PVDF membranes that do not bind Ponceau S well are used, UV trans-illuminator can be employed to visualize proteins after transfer.

2.4 BLOCKING THE MEMBRANE

Following transfer, Ponceau S staining and cutting/trimming, membranes need to be subjected to a "blocking" step, in order to decrease nonspecific binding of primary/secondary antibodies. Blocking solutions contain proteins which bind to the "sticky" sites on the membranes, thus preventing the antibodies from binding nonspecifically to those sites and thereby, reducing the background/noise.

The most popular blocking solutions for are 5%−10% nonfat milk or 3−5% BSA (fraction V) dissolved in TBST. Although milk is cheaper, it is a complex solution and may not the best choice for certain antibodies. For example, milk contains phospho-casein, which may interfere with certain phospho-specific antibodies, thus, increasing the background/noise and/or decreasing the signal. Particularly, milk should not be used when streptavidin-biotin detection is used, as milk contains biotin and will interfere with the detection.

BSA is a safe blocking agent for most phospho-specific antibodies. One disadvantage of BSA is that it cannot be used when lectin probes are used as BSA contains carbohydrates. For non-apparent reasons antibodies may perform better in one blocking solution versus the other, and therefore, it is important to follow the antibody manufacturer's recommendations. Incubation time for blocking is typically 1 hour at RT on a shaker, followed by rinses with TBST to remove the excess blocking solution prior to incubation with the antibody solution. For most antibodies, the time can be reduced to 30 min and prolonged blocking for up to 2 hours should not negatively affect the results. Several membranes (up to 3−6, depending on the size) can be blocked in one box.

2.5 PRIMARY ANTIBODIES

Antibodies, also known as immunoglobulins (Ig), are naturally produced by the immune system in response to antigens, such as microorganisms. Primary antibodies for Western blotting are often produced in animal hosts such as rabbit, mouse, rat, goat and sheep, and occasionally in llama and sharks. Primary antibodies can be developed to not only recognize a protein but also specific modification, such as phosphorylation, acetylation, glycosylation, or methylation.

Antibodies are Y-shaped molecules and they are comprised of two copies of identical heavy chains and two copies of identical light chains (Fig. 2.4). The chains are connected by disulfide (−S−S−) bonds and reducing agents such as β-mercaptoethanol or dithiothreitol are

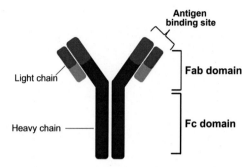

Figure 2.4 Diagram of a typical IgG antibody structure.

required to fully denature antibodies. The upper part of the Y-shape of an antibody is comprised of two antigen-binding fragments or F(ab) regions and the lower part is called "crystallizable" or "constant" region (Fc region). Primary antibodies can be covalently conjugated to various probes such as biotin of digoxigenin (DIG), fluorescent dyes, enzymes, as well as agarose or Sepharose, via conjugation to Fc-binding bacterial proteins called protein A and protein G (see Appendix K).

Monoclonal antibodies are an ideal tool for Western blotting since highly-specific and high-affinity clones are usually selected during the antibody generation process. Moreover, they are relatively easy to produce at high quantities, providing a reliable source of a consistent high-quality reagent, with little to no variation between batches. However, generation of monoclonal antibodies is an involved process.

Polyclonal antibodies are cheaper, easier, and quicker to generate. Moreover, they provide an extra signal amplification step, since more than one antibody molecule from a polyclonal pool will bind per protein. This further enhances sensitivity of Western blotting. Moreover, polyclonal antibodies are usually better reagents for immunoprecipitation. However, not only each host animal will a different quality polyclonal antibody batch, but also different serum bleeds from the same host will have a varying level of quality. The disadvantage of polyclonal antibodies is thus, a limiting amount that can be obtained from each batch and a significant batch-to-batch variability.

Although antibodies are a valuable tool, they need to be optimized and validated before use. Usually, this is done by the antibody

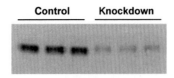

Figure 2.5 An example of a total antibody validation by knockdown.

manufacturer. However, it is always important to ensure that the obtained signal is specific. The best practice is to knockdown or knockout the target protein and run such negative control lysates on each gel (Fig. 2.5). If dealing with a phospho-specific antibody, a nonphosphorylatable mutant (i.e., Ser/Ala, Thr/Ala, or Tyr/Phe) should be used as a negative control.

Membranes are incubated with primary antibodies either overnight at 4°C or for 1−2 hours at RT, and this depends on the binding affinity of the antibody for the target protein and the abundance of the target protein. When using polyclonal antibodies, multiple primary antibodies bind per target protein, providing a crucial signal amplification step, contributing to the sensitivity of Western blotting.

The final working concentration for antibodies should be used as indicated by the manufacturer. However, many times, even 2−3-fold lower concentrations work too, as the optimal antibody concentration depends on the cell type and amount of protein loaded. Too high antibody concentration can result in high background and too little antibody will give faint bands. If the dilution information for an antibody is not available, start with 0.5−1 µg/mL concentration and then dilute (if necessary) to optimize the signal.

Incubation of antibodies at 4°C can increase the specific binding of antibody to the target protein and lower the background noise, whereas incubation at 37°C will shorten the necessary incubation time to 30 minutes for total antibodies.

Primary antibodies are best diluted in 3%−5% BSA in TBST. However, certain antibodies may require to be diluted in milk, in order to reduce nonspecific background. We recommend to supplement all primary antibody solutions with 0.05% sodium azide, in order to inhibit microorganism growth and prolong the reusability of the antibody solutions (Fig. 2.6).

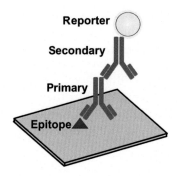

Figure 2.6 A simplistic diagram of the ternary complex between an epitope, a primary antibody and a secondary antibody conjugated to a reporter.

2.6 SECONDARY ANTIBODIES

The secondary antibody dilution may range from 1:1000 to 1:20,000, depending on the brand, target protein abundance and primary antibody affinity. Secondary antibodies can be diluted in TBST or blocking buffer. Although diluting secondary antibodies in the blocking buffer can result in lower background noise, this also might give a weaker signal because the blocking protein can interfere with the binding of the secondary antibody to the primary antibody. Secondary antibodies are usually incubated for 1 h at RT, followed by three washes with TBST to remove residual antibodies from membranes prior to the developing process.

Multiple secondary antibodies bind per primary antibody, providing a crucial signal amplification step, contributing to the sensitivity of Western blotting. Secondary antibodies are usually conjugated either to an enzyme such as horseradish peroxidase (HRP), alkaline phosphatases (AP), or to fluorescent probes. Although the enzyme-conjugated secondary antibodies amplify the signal and further enhance the sensitivity of Western blotting, signals from fluorescent probe-conjugated secondary antibodies can be quantified more reliably.

It is best to avoid reusing secondary antibodies. Moreover, never supplement secondary antibody solutions containing sodium azide, which will completely inhibit HRP and result in a complete loss of signal.

Different antibodies require different storage; therefore, it is important to check the datasheet provided by the antibody manufacturer for proper storage conditions. Some antibodies require storage at $-20°C$

or $-80°C$, whereas others need to be kept at $4°C$. Some antibodies do not perform after freeze–thaw cycles; therefore, it is important to aliquot into single-use volumes to avoid the freeze–thaw process.

A rule of thumb is that antibodies that are conjugated to enzymes should not be frozen, as the activity of the enzyme will decline sharply after a freeze–thaw. Antibodies that are conjugated to fluorophores should be kept in dark during storage and during experiments, in order to avoid photo-bleaching.

2.7 BLOT WASHES

After incubation with the primary/secondary antibody, the membranes can be washed 3–4 times with TBST for 5–10 minutes each wash, in order to remove the unbound primary antibody before incubating with the secondary antibody and wash off secondary antibody before adding the enhanced chemiluminescence (ECL) reagent or proceeding with any alternative detection method. Side-to-side shakers/rockers are recommended over radial shakers. About 15–20 mL of TBST per mini-gel membrane is usually sufficient. Several membranes (up to 3–6, depending on the size) can be washed in one box, provided they were incubated with the same antibody. By increasing the speed of the shaker, the time necessary to complete the washes may decrease. The standard TBST formulation (50 mM Tris–HCl pH7.5, 150 mM NaCl, 0.1% Tween 20) is stringent enough for most antibodies. However, the NaCl and Tween 20 concentrations can be increased 1.5–2-fold in order to decrease background and increase specificity.

2.8 DEVELOPING WESTERN BLOTS

Two popular methods of used detection are the ECL and the direct fluorescence systems (e.g., Odyssey Imaging Systems—LI-COR). One of the most important advantages of using ECL is that it yields higher sensitivity when compared to fluorescent systems, and this can be important aspect if the abundance of the target protein is low or the primary antibody does not have a high affinity for the molecule of interest.

One of the most common chemiluminescent detection is based on a chemical reaction where HRP enzyme in the presence of a peroxide buffer catalyzes the oxidation of luminol and results in emission of visible light. AP is also able to catalyze the chemiluminescent reaction.

As secondary antibodies are complexed with the primary antibodies, which are bound to the protein of interest on the membrane, the amount of the light produced by the ECL reaction is directly proportional to the amount of the epitope on the membrane. The emitted light be detected by exposure to a light-sensitive X-ray film or can be captured using a charge-coupled device (CCD) camera. For most ECL reagents, the intensity of light peaks shortly after 5 minutes and strongly decays after 25−30 minutes; therefore, the membranes need to be exposed right after the incubation with the ECL reagents. Multiple film exposures should be done, in order to obtain an exposure with the best signal to noise ratio. The X-ray films usually provide a more sensitive solution to signal detection, than most modern CCD devices and fluorescence readers.

Fluorescent Western blotting (e.g., LI-COR) offers the ability for increased quantification accuracy and a convenient way for multiplex detection, where multiple protein of interest can be detected simultaneously. Fluorescence is emitted when a fluorophore is excited at a certain wavelength, and unlike the enzyme-driven light emission, occurring during the ECL method, there is no amplification of the signal. This lack of amplification gives the method its quantitative nature, due the fact that the signal stays in a linear range. On the other hand, the lack of signal amplification results in a relatively lower sensitivity of the method, compared to the ECL method.

For fluorescent detection secondary antibodies are labeled with near-infrared fluorescent dyes. The fluorescent signals are stable over months and the fluorescent intensity is not altered by exposure time yielding consistent and quantitative data. This method of detection is time-saving and could offer reduced costs compared to the chemiluminescent detection system.

CHAPTER 3

Good Practices

Take your time and do it right.

—*Nick Fury*

Can't afford to chance it.

—*Rick Sanchez*

I'm not naive to what it takes.

—*James Adams*

Good Practice #1: Make sure that the glasses and combs are very clean

Most of air bubbles and other gel imperfections come from the dust and acrylamide particles left on the glass surface. Store dirty glasses in distilled/deionized water right after you are done using them. This way, you will not need to scrub/soap them too much next time you will need to wash them. The easiest way to wash glasses stored in this way is to use hot water and a piece of tissue paper or sponge to scrub

Western Blotting Guru. DOI: http://dx.doi.org/10.1016/B978-0-12-813537-2.00003-5

the particles off. No soaping is usually required. After hot water, rinse the glasses with distilled/deionized water and air dry.

Good Practice #2: Always keep the running and transfer tanks clean

Immediately after run/transfer is finished, rinse the tanks once with hot water and twice with distilled/deionized water before the buffers can dry out. Keep the lids on, so that dust doesn't get inside the tanks. Do not leave the tanks with the buffers inside them, as it will corrode and affect the electrode life-span. There is no need to completely dry out the tanks before storing them.

Good Practice #3: Always aim for loading replicates

If you have enough wells, aim for loading the same sample at least twice, in order to obtain a technical/loading replicate and get a more reliable readout of the data. Western blotting data can be occasionally obscured by bubbles and other technical issues from the transfer step, and therefore, duplicates usually increase reliability of the obtained data.

Good Practice #4: Make sure that the membranes and papers are clean

A simple way to organize nitrocellulose/polyvinylidene fluoride (PVDF) membranes and the blotting papers is to perform a large-scale cutting and keep them in a ready-to-go precut format in a clean box. Use clean surfaces to cut. Do not leave marker/pen traces or indentations on them. Always wear clean gloves and pick up the membranes by their corners, when handling them. Make sure that the buffers in which they are pre-wet are clean and are placed in clean containers.

Good Practice #5: Always prepare the transfer buffer fresh

It is best to store $10 \times$ transfer buffer at room temperature (RT) and prepare $1 \times$ buffer on the day of the transfer. Determine an optimum volume needed for a routine and always prepare just as much as needed to avoid wasting ethanol/methanol. Buffer that is used for pre-wetting the membranes and sponges can also be poured into the transfer tank, once the cassettes are assembled, in order to minimize the transfer buffer waste.

Good Practice #6: Always prechill the transfer buffer

Addition of ethanol/methanol to the $10 \times$ buffer results in a noticeable increase of the final $1 \times$ buffer temperature. It is therefore best to pre-chill the ready 1 × transfer buffer by placing the bottle containing it into ice-water bath for 30 minutes. Alternatively, the bottle can be placed into the cold room or fridge for 1−2 hours.

Good Practice #7: Ponceau S quality check

It's important to know that the run/transfer quality was acceptable before blocking and incubating the blots with the antibodies. The easiest way to check this, is to incubate the blots right after the transfer in 0.1% Ponceau S solution in 5% acetic acid for 30 seconds. Collect and reuse Ponceau S indefinitely and only top-up the used solution with new solution. There is never any need to throw away the used solution as the dye is very potent. Membranes can then be rinsed with distilled water 2−3 times and blots can be quality checked and cut into smaller parts/slices, if necessary to incubate with different antibodies. Ponceau S will wash away once the blots are immersed into milk, but its best to briefly rinse them with Tris-Buffered Saline Tween 20 (TBST) before blocking in milk. Ponceau S staining will also give a quick assessment of how equal the lanes were loaded and can be used as an internal control, as well as protein purification quality check. Ponceau S staining of nitrocellulose membranes is a more sensitive method of total protein visualization than Coomassie staining of gels (Figure 3.1).

Good Practice #8: Always load a positive and a negative control for antibody specificity

Obviously, experiments have to be designed with positive and negative controls, in order to be able to correctly interpret the data.

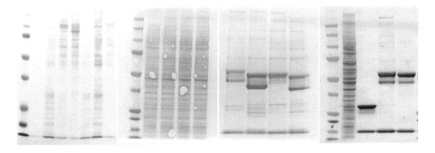

Figure 3.1 *Examples of membranes stained with Ponceau S to visualize proteins after transfer.* Ponceau S staining can provide valuable information about internal loading consistency, run/transfer quality, as well as protein purity. First panel—different fractions from a gel filtration elution. Second panel—a membrane with multiple bubbles, indicating problem with transfer. Third panel—four eluted purified proteins show contaminating proteins. Fourth panel—Lysate lane, control GST pull-down lane, and two lanes showing GST pull-down of a target protein in the presence and absence of a treatment.

In addition, if the experimental design doesn't already include a positive and a negative control for antibody specificity, load such controls after loading the experimental samples. Prepare single-use aliquots of the positive and negative control samples for antibody specificity ready to be heat-denatured and loaded. Such controls become even more critical when antibodies and especially, phospho-specific antibodies are reused several times.

Examples: (1) Cell lysates from untreated and drug-treated cells; (2) purified protein; and (3) purified wild-type and Ser/Ala mutant protein (for phospho-specific antibody specificity—especially when reusing polyclonal phospho-specific antibodies).

Good Practice #9: Always use fresh APS

Prepare 50 mL of 10% APS, filter through a 0.45-μm syringe filter and aliquot into single-use volumes. Store at $-20°C$ for up to 3 months. Avoid storing in autodefrost freezers.

Good Practice #10: Avoid autodefrost freezers

Some labs do not use autodefrost freezers, whereas some prefer them for certain applications. Avoid storing your samples (in $1 \times$ sodium dodecyl sulfate polyacrylamide gel electrophoresis (SDS-PAGE) sample buffer) in autodefrost freezers, as the cyclic temperature fluctuations in such freezers may negatively affect the protein stability and therefore the sample quality.

Good Practice #11: Always use fresh enhanced chemiluminescence (ECL) solutions and fresh H_2O_2

Home-made ECL solutions should be used up within 2 weeks of preparation since the signal intensity will start to fade. The H_2O_2 for the home-made ECL preparation has to be very fresh. Most solutions of H_2O_2 (even with stabilizer) are only good for about 6 months, when stored at $-20°C$. Buy smaller volumes of H_2O_2 every 6 months to ensure high-quality home-made ECL reagent. The luminol and p-coumaric acid solutions should be single-use aliquoted and stored at $-20°C$.

Good Practice #12: Always use clean/fresh clear plastic file pockets in the film cassette

Either wash the plastic pocket with water and dry it out completely before reusing or use a fresh one. The dried out salts and water can interfere with the ECL reaction and affect the results. Also, don't let ECL solution from previous sessions to accumulate behind the plastic pocket.

Good Practice #13: Use an optimum number of films for exposure

X-ray films are expensive and, unless you recycle them, are environmentally unfriendly. When using multiple blots with different antibodies in a single film cassette the best routine for exposures is to expose for 3, 10, 20, and 30 seconds and then for 1 and 2 minutes. Depending on the results a very short exposure (1 s) and a longer exposure (10 min) can be taken.

Good Practice #14: Allow enough space for membranes in the incubation boxes

Cut the membranes and choose box sizes in such a way that the membranes can freely shake inside the boxes. This is especially important for wash and secondary antibody steps, and when using multiple membranes in the same box. Enough room for proper shaking of the membranes will prevent them from sticking to each other and therefore will prevent uneven antibody binding and incomplete washout of excess antibody.

Good Practice #15: Use fresh TBST

Always make sure that TBST doesn't have any contamination in it. Prepare volumes of $1 \times$ TBST that can be used up in less than 1 week. If preparing higher volumes, store them at 4°C/cold room.

Good Practice #16: Dry the membranes properly

Don't let the membranes to dry out inside the film cassette. Rinse them with water to remove the ECL and dry afterward. Drying membranes with ECL or TBST will allow crystals of salt to form and this will damage the membrane.

Good Practice #17: Filter bovine serum albumin (BSA)

Filter 5% BSA in TBST + 0.05% NaN_3 (sodium azide) stock through a 0.45-μm bottle-top filter. Such filtering step will eliminate any "dust", hair and other particles that can stick to the membrane during primary antibody incubation and will ensure that you don't get spots or dots on your membrane. Avoid using NaN_3 in TBST solutions or solutions in which horse radish peroxidase (HRP)-conjugated secondary antibodies will be diluted, as NaN_3 is a potent HRP inhibitor and will result in no ECL signal.

Good Practice #18: Ensure the transfer cassette thickness is optimal

Over time, the sponges get compressed and old sponges tend to get thinner than new ones. Adjusting the number of thin blotting papers in a transfer cassette in order to compensate for compression will ensure that the gels are not squeezed too hard or too little. Not enough squeezing of the gels can lead to increased bubbles and bad transfer efficiency/uniformity, whereas too much squeezing of the gels can lead to deformation of the bands or even breakage of the gels.

Good Practice #19: Prewet membranes and blotting papers properly

Nitrocellulose membranes should be pre-wet for 1−3 minutes in transfer buffer. PVDF membranes need ~ 1 minute in methanol/ethanol, before they can be pre-wet in transfer buffer for 3−5 minutes. Blotting papers should be pre-wet in transfer buffer for 15−20 minutes. Shorter than recommended pre-wetting can lead to low quality transfer and bubble formation during transfer. Pre-wetting the blotting papers for too long will make them become too soft and hard to handle and they will shed too many particulate fragments that can decrease transfer quality. Pre-wetting membranes for too long is also not recommended.

Good Practice #20: Verify stripping

For bands that are close together by size or if it is a phospho-then-strip-then-total reprobe you plan to do, it is always critical to verify that the stripping worked completely by incubating the blot with the relevant secondary antibody and developing it. This is because occasionally, residual leftover "unstripped" antibody can still give a significant signal and cause interference with the data analysis. This is particularly important if the quick stripping protocol (Appendix M) is used.

Good Practice #21: Not so fresh buffers

For most applications, use fresh buffers and avoid reusing running or transfer buffers, as the buffering capacity will decrease after the run/transfer and proteins will be present in the used buffers. Transfer buffers can be occasionally reused two to three times when medium−small molecular weight (MW) proteins are used. However, this will affect transfer efficiency reproducibility between immuno-blotting sessions.

Good Practice #22: Proper running tanks

One of the biggest losses of time and effort happen when the running tanks do not perform properly. All kinds of "smiles" and streaks can happen due to a broken running tank cell. It is obvious that such tanks need to be repaired/replaced; however, many times common equipment is not sent for repair/replacement.

If you're not sure about the quality of the running tank at hand, a brief (10−15 seconds) pre-run can be given before loading any samples, in order to see if the tank runs at all by monitoring bubbles coming out of the electrodes, in order to see if the cell is functional or not.

The intensity of bubbles released from the electrode is proportionate to the current passing through the cell and if one of the cells is not performing properly, there will be either too few or too many bubbles.

Good Practice #23: Nitrocellulose vs PVDF

Keep in mind that some hydrophobic proteins bind to PVDF a lot better than to nitrocellulose and choice of the right membrane type might result in a big difference in the data quality. For example, lipidated LC3 binds to PVDF much better than to nitrocellulose.

Good Practice #24: Don't overheat

The $5 \times$ sample buffer needs to be slightly heated up at 70−90°C for 1−2 minutes to dissolve the frozen SDS. However, it is best to minimize the length of heating required to thaw buffer and not leave the buffer in the heat block for longer than needed to thaw. Also, don't heat your proteins for longer than 10 minutes, as this will affect the sample quality.

Good Practice #25: Immerse in ECL

The best way to ensure that ECL substrate reaches all parts of the membrane uniformly is to immerse the membrane into the ECL solutions and gently shake for 1−2 minutes. 15 ml of ECL solution is usually enough for 5−6 mini-gel membranes in the same box. Many commercial ECL solutions work even if diluted 2−5-fold. Thus, determine at what dilution you can get a reliable signal and always immerse the membranes in the ECL solution, instead of attempting to cover the membrane surface with a low volume of concentrated ECL solution, which may be tricky and unreliable.

Good Practice #26: Load the wells evenly

Load empty wells with $1\times$ sample buffer, so that the lanes are evenly loaded. If you are running cell lysates and immunoprecipitation samples on the same gel, load $1\times$ sample buffer between them, as even a minute amount of cell lysate can significantly contaminate the immunoprecipitation lane. Similarly, the bait protein signal from the immunoprecipitation lane is usually much stronger than the one from the cell lysates and that may contaminate the cell lysate lane making it appear to have more bait input.

Good Practice #27: Do not overload wells

Loading wells to their maximum capacity can cause overflow of the sample to the next well and affect the outcome of the experiment. It is always better to load up to 75% of the well volume, in order avoid accidental overflow.

Good Practice #28: Do not load too much of the protein marker

Protein markers tend to have a higher density then the protein sample loaded and this can affect the running of the sample next to it. If you can spare an extra lane, load $1\times$ of sample dye between the marker and the sample.

Good Practice #29: Remove the acrylamide from the wells before running the samples

Aspirate the wells or wash them with water to remove potential acrylamide debris after removing the combs. This can be done in the sink, or after assembling the glasses into the running tanks.

Good Practice #30: Take care of the electrodes

The electrode tips in the running and transfer tanks are relatively fragile. After washing the tanks do not dry the tanks upside down with their weight over the electrode tips as this will result in loosened or even broken running/transfer cells.

CHAPTER 4

Optimization and Troubleshooting

A person who never made a mistake never tried anything new.

—Albert Einstein

It's a stupid problem to have, but it's a problem nonetheless.

—Hector Tapia

When obstacles arise, you change your direction to reach your goal; you do not change your decision to get there.

—Zig Ziglar

4.1 OPTIMIZATION RULES

Most antibodies provided by the commercial vendors fit a certain quality criteria that do not require any significant optimization to be done by the user. However, both user-generated and reagent-dependent problems can arise (Figure 4.1) and depending on the target sample type (e.g., cell lysate vs tissue lysate), or when in-house-generated antibodies are used,

Western Blotting Guru. DOI: http://dx.doi.org/10.1016/B978-0-12-813537-2.00004-7

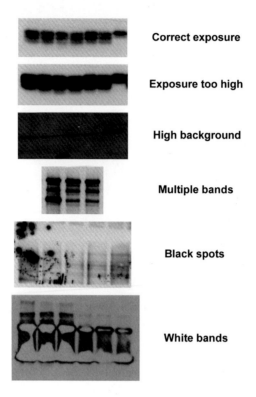

Correct exposure

Exposure too high

High background

Multiple bands

Black spots

White bands

Figure 4.1 Examples of different types of western blotting problems.

several optimization strategies may need to be undertaken, before reliable, specific and reproducible and optimal protocols for blocking method, final antibody concentration, washing steps and even total protein loading levels are achieved.

The rules of the western blotting optimization game are such: the higher the amount of protein loaded, transfer time, primary and secondary antibody concentration, incubation time, exposure time, the higher will be the signal, but so will be the background. Thus, it is critical to find the middle point, where the amount of signal is strong enough to analyze the data (i.e., strong shades of gray), while the background is low enough that the data is easily analyzable, specific and is of publication quality.

4.2 GENERAL OPTIMIZATION STRATEGIES

4.2.1 Antibody Concentration
Optimum working concentrations of antibodies are generally determined empirically, as the affinity and specificity of antibodies varies,

as does the abundance of the target epitope in a given sample. This dilution information is usually provided by the vendor and changes depending on the application. Most antibodies are sold at concentrations high enough for the final dilution for immunoblotting to be 1:1000. This results in $0.1-1$ μg/mL final concentration, depending on the antibody. Most antibodies will work well at around 0.5 μg/mL concentration; however, this depends on several factors, such as the amount of protein loaded into gel, transfer efficiency, antibody incubation time, extent of washing, secondary antibody dilution, and epitope abundance.

Therefore, it is possible to save on antibody amount used by considering and adjusting these factors. If an antibody is raised in-house, optimum final antibody concentration can be determined by testing it at 0.1, 0.3, 0.6, 1, and 2 μg/mL final concentrations on a membrane with a positive control. It is sometimes possible to use lower antibody concentrations for overnight incubations and use higher concentrations to shorten the incubation time. After a certain point, higher concentration of antibody usually leads to nonspecific detection. Therefore, when antibody concentrations are increased to increase signal yield and shorten incubation time, maximum allowed concentration at which the antibody is still specific has to be known.

Within limits, antibody concentration is directly proportional to sensitivity and signal strength, and inversely proportional to specificity and required incubation time.

[Antibody] ⬆ → Sensitivity, Signal strength ⬆ & Specificity, required incubation time ⬇

4.2.2 Antibody Incubation Times

Most "total" antibodies that are not raised against a tricky epitope such as a phosphopeptide will bind to their antigens quickly enough and immunoblotting incubation times can therefore be kept at around $0.5-1$ hour at room temperature.

However, some weak "total" and most phospho-specific antibodies require a longer incubation (usually overnight) and therefore these incubations should be kept at 4°C. Moreover, if the antigen is not very abundant, antibody−antigen saturation may need to be high and therefore, overnight incubation at 4°C may be required.

On the other hand, certain phospho-specific antibodies have a very high affinity for their target epitope and/or the epitope abundance is high enough to perform 1–2-hour-long incubations at room temperature to obtain satisfactory results. Such information is usually not provided by the vendors and therefore needs to be determined empirically for the system (i.e., cell line, lysate type, treatment) the experimenter is employing.

Within limits, incubation time is directly proportional to sensitivity and signal strength; and inversely proportional to required antibody concentration. However, after a certain length of incubation time the specificity might decrease due to oversaturation.

Incubation time ⬆ → Sensitivity, Signal strength ⬆ & Specificity, [antibody] ⬇

4.2.3 Wash Stringency

The washing times can usually be increased from three to six times, which will increase the signal/noise ratio, but may also decrease the overall signal. However, for certain difficult case antibodies to be specific enough and/or to give a good enough signal/noise ratio, the wash stringency needs to be increased too. To change the stringency of the washes, the concentrations of either NaCl and/or Tween-20 in the TBST can be optimized. NaCl can be increased from 150 mM to 200, 250 or even 500 mM. Tween-20 concentration can be increased from 0.05% to 0.1% or 0.2%. In rare cases, low concentrations of a stronger detergent such as Triton X-100 (0.1–0.5%) or even SDS (0.02–0.1%) can be supplemented to get rid of the background.

4.2.4 Blocking

5%–10% nonfat milk in TBST blocking for 1 hour at room temperature (RT) is a proven cheap, strong and reliable method for blocking nonspecific signals from most antibodies. However, some antibodies are not compatible with milk, as their epitopes can be present in milk proteins. In such cases, milk is substituted for BSA or casein blocking. In rare cases, donkey serum can also be used to block. Overblocking is a potential problem that can increase the background, so blocking for longer than 1 hour at RT is not recommended. Many antibodies will give a good signal/noise ratio with even 5% or even 3% milk as a blocking agent, when blocked for 1 hour, or even 30 minute at RT.

It is best to prepare the blocking milk solutions fresh, on the day of use, but when supplemented with 0.05% sodium azide, they can be kept at 4°C for up to a week.

4.2.5 Preclear the Antibodies

One of the simplest methods to increase antibody specificity is to perform a preclearing step by incubation with a membrane that contains the whole proteome, except the target protein. Knockout or knockdown lysate therefore can be used to perform such negative selection for a total antibody. For phospho-specific antibodies, bacterially-expressed purified wild-type proteins, Ser/Ala (or Tyr/Phe) mutants or dephosphorylated proteins can be used to allow removal of antibody species that detect the nonphosphorylated form of the target protein from the antibody solution. An overnight incubation is recommended to saturate the membrane with such nonspecific antibody species and allow only the specific species to be left in the solution, as much as possible. For some cases, multiple rounds of such negative selection can be applied.

4.2.6 Supplement Antibody Solutions With Nonphosphorylated Peptides

If a phospho-specific antibody partially crossreacts with the nonphosphorylated version of the epitope, the easiest (but not the cheapest) way to get rid of the background signal is to supplement the antibody solution with the nonphosphorylated peptide at tenfold molar excess. In other words, add 10 μg/mL of nonphosphorylated peptide (phosphorylated version of which the antibody was raised against) to the antibody solution of 1 μg/mL and incubate this mixture with the membrane. The peptide will sequester the nonspecific species of the antibody in solution and prevent it from binding to the nonphosphorylated protein on the membrane, thereby, increasing the specificity of the obtained signal.

4.3 TROUBLESHOOTING SPECIFIC PROBLEMS

4.3.1 Problem Type 1

No bands were observed or the bands were weak/faint.

The easiest way to figure out the source of the problem of this type is to always include a positive control (e.g., cell lysate or purified protein) in every gel. Dedicate a lane for positive controls.

4.3.1.1 Subtype 1
Proteins were detectable on the membrane by Ponceau S, but no bands were observed or the bands were weak/faint after the enhanced chemiluminescence (ECL) step

Potential Problems	Potential Solutions
Wrong secondary antibody was used	Check whether the primary was raised in rabbit or mouse, for example, and use the correct secondary antibody
Primary was used against wrong target protein species	Many antibodies are raised against human epitopes and do not crossreact with proteins from other species, such as rodents. And vice-versa. Check if the antibody manufacturer shows evidence that the antibody you are using is supposed to detect proteins from the species you are using
Target protein is not expressed in the cells/tissue	Check which samples/lysates has the antibody manufacturer provided as a positive control. Load them on the same gel with your samples to rule in/out this potential problem
Target epitope is hindered by a posttranslational modification, (e.g., phosphorylation)	• Check which samples/lysates has the antibody manufacturer provided as a positive control. Load them on the same gel with your samples to rule in/out this potential problem • Check www.phosphosite.org for phosphorylation reports of the peptide against which the antibody was raised • Use Lambda phosphatase or Calf-intestinal phosphatase to dephosphorylate your protein/lysates (see Appendix G)
Target epitope is cleaved off during treatment	Use a second antibody raised against a different region of the protein. Run a tagged-version of the protein as a positive control
Target epitope is missing due to alternative splicing	Prepare total cDNA from the target cell line or tissue. Sequence the region where the epitope is located to confirm that the exon containing the epitope is present in the splicing variant expressed in the target cell line or tissue
Primary/secondary antibody was used at a wrong dilution	Use final concentrations recommended by the antibody manufacturer. Use $2\times$ higher concentration of the primary/secondary antibody
Primary antibody dilution is old or doesn't survive freeze−thaw	Prepare fresh antibody dilution. Store antibodies that do not survive freeze−thaw well at 4°C
Phosphosite is not phosphorylated due to lack of correct treatment conditions	Identify which treatment conditions are required to induce the target phosphorylation
Expired primary antibodies, secondary antibodies, ECL or developer reagents were used	Always use fresh reagents. Especially, H_2O_2 added to home-made ECL should be very fresh
The membrane was cut wrongly and the bands were cut out	Never cut the membranes inside the region that contains proteins. This practice will also ensure that unexpected unknowns are not missed. (e.g., unexpected cleavage during a treatment)

(Continued)

(Continued)	
Potential Problems	**Potential Solutions**
The target protein was not denatured properly	Some epitopes will not be revealed if the protein samples are not heated well enough. Try heating for 5 min at 95°C or for 10 min at 70°C
Incubation time was not long enough	Incubate the blots with the primary antibodies according to the manufacturer's recommendations. Total protein antibodies usually work when incubated for 1 h at RT, but give more specific results when incubated for 16 h at 4°C. Phospho-specific antibodies usually required the overnight incubation, but some very strong antibodies can be incubated for 2–3 h at RT. Occasionally, 48–72 h incubations at 4°C give better results
Incubation with the ECL reagent was too long or too short	Use ECL according to the manufacturer's recommendations or as outlined in Appendix E, if using home-made ECL
Exposure time too short	Increase the exposure time by 2–3-fold
ECL reaction has finished	Make sure that the membranes are exposed to the films during the first 20–30 min for (most) commercial ECLs and within the first 10–15 min for home-made ECL (see Appendix E)
The membrane was stripped and reprobed too many times or over-stripped	Rerun the samples. Always reprobe with the strongest antibody last. Use quick stripping protocol (Appendix F) if a membrane is going to be stripped many times, as it is a milder stripping method
Primary/secondary antibody was not properly stored	Follow manufacturer's storage recommendations. Some commercial antibody solutions cannot be frozen. Others need to be aliquoted into single-use aliquots. Store antibody only in manual defrost −20°C freezers or −80°C freezers

4.3.1.2 Subtype 2
Proteins were *not* detectable on the membrane by Ponceau S

Potential Problems	**Potential Solutions**
Not enough protein was loaded or transfer was incomplete	Some proteins are not abundant and if low amounts of protein are loaded or the transfer is not done long enough (especially for proteins with high MW) and combined with other suboptimal conditions may lead to no bands. Load at least 20 µg of total protein lysate or 200 ng of purified protein. This can be increased to up to 100 µg for lysate sand 0.5 µg for purified proteins. Transfer at least for 1 h at 100 V. This can be increased to 2 h at 100 V
Proteins were over-transferred	For small proteins, this can be a problem. Decrease transfer time to 0.5 h at 100 V. Use membranes with smaller pore size. Use gels with higher concentrations
Ponceau S stock is old or has been reused too many times	Prepare fresh Ponceau S stock

(Continued)

(Continued)	
Potential Problems	**Potential Solutions**
Proteins were degraded and/or dephosphorylated prior to loading	Store protein samples in SDS–PAGE sample buffer at −20°C and avoid repeated freeze–thaw cycles. Prepare fresh samples. Do not leave samples on bench top for prolonged time periods. Do not heat protein samples for longer than 10 min
Proteins were transferred in the wrong direction	Always include a prestained MW marker and stain with Ponceau S to rule out this problem. If the marker is not visible on the membrane either, the transfer may have been done in the wrong direction

4.3.2 Problem Type 2
Multiple bands were observed (overall background nominal)

Potential Problems	**Potential Solutions**
Too much protein loaded per lane	Decrease protein amount loaded per lane by 2–3-fold. Loading more than 20 μg total protein per lane will result in nonspecific bands. Use protein quantification assays such as Bradford or BCA assay
Too much primary/secondary antibody was used	Decrease the antibody concentration by 2–3-fold. Decrease incubation time. Decrease incubation temperature. Overnight incubations at 4°C usually give more specific results
The antibody is not specific enough	Preclear the antibody with a membrane containing lysates of cells with a knockout or knockdown for the target protein. See Section 4.2.5. Obtain a better commercially available antibody
The target protein is either cleaved, degraded or underwent posttranslation modifications, such as phosphorylation, ubiquitination, SUMOylation, NEDDylation	Use deubiquitinase/phosphatase inhibitor-free lysis buffer to allow the endogenous enzymes remove the posttranslational modifications (during lysis) to confirm this. Use more protease inhibitors in the lysis buffer or lyze in 1 × SDS–PAGE sample buffer to inhibit the proteases and prevent target protein degradation. Perform more specific experiments to address the more likely scenario, depending on the MW of the additional bands
The exposure is too high	At very high exposures, even the best antibodies will give off-target bands. Decrease exposure time by 2–3-fold
Incubation time too long	Decrease incubation time to 1 h for total antibodies and 16 h for phospho-specific antibodies. Decrease incubation temperature
Heavy/light antibody chains from immunoprecipitation were detected	If the target protein band is too close to the heavy/light chains of the antibody used for immunoprecipitation, follow recommendations #8 and/or #9 in Chapter 5, Tips and Tricks

4.3.3 Problem Type 3
Band smears were observed

Potential Problems	Potential Solutions
Low quality sample—cell debris in the sample	Spin sample at 16,000 × g for 1–3 min to sediment the insoluble particles
Low quality sample—immunoprecipitation beads in the sample	Spin sample at 16,000 × g for 1–3 min to sediment the beads. Use Spin-X columns to remove beads from the samples
Low quality sample—too much unsheared genomic DNA in the sample	If the sample is too viscous, sonicate it to shear the genomic DNA. Benzonase can be used instead of sonication
Low quality sample—too much glycerol or detergent in the sample	Dilute 2–3-fold if the final detergent or glycerol concentration is too high in the sample
Low quality gel—acrylamide particles in the well	Rerun on a new gel. Rinse the wells before loading
Low quality gel—bubble/particle in the gel	Rerun on a new gel. Check gels for bubbles before running. Clean the glasses to remove particles and avoid bubble formation
Low quality gel—expired gel	Adhere to expiration dates for commercial gels and store home-made gels for less than 2 weeks. See Chapter 5, Tips and Tricks, #23
Too much protein loaded	Overloading the lanes oftentimes results in smears. Load 2–3-times less total protein
Gel run was done at very high voltage/current	Follow recommended voltage/current ranges provided by the gel tank manufacturer. Decrease voltage/current by 1.5-fold. Let the samples enter and pass through the stacking gels at low voltage (∼70 V)
Heavy/light antibody chains from immunoprecipitation result in smears	If the smears originate from the 50-kDa/25-kDa heavy/light chains of the antibody used for immunoprecipitation, follow recommendations #8 and/or #9 in Chapter 5, Tips and Tricks. Load threefold less sample
Old buffer was used	Always use fresh running buffers
Protein was not denatured properly	Some proteins will form insoluble oligomers or migrate as smears if denatured too much or not enough. Use fresh 5× SDS–PAGE sample buffer. Denature samples at either 70°C for 10 min or 90–95°C for 5–10 min
Protein is glycosylated and/or is a transmembrane protein	Glycosylated and/or transmembrane proteins many times migrate as smears. Try not heating the samples or heating at 55–70°C for 10 min, instead of 90–95°C
Protein is heavily posttranslationally modified (e.g., ubiquitination, phosphorylation)	Load 1.5–2-fold less protein

4.3.4 Problem Type 4
High overall background

Potential Problems	Potential Solutions
Blocking was not done properly	Use 5%–10% fresh milk in TBST to block for 30 min–1 h at RT. Use only nonfat milk. Do not block for excessively longer times
Blocking reagent is inadequate	Some antibodies will bind to proteins in the milk. Use 5% BSA in TBST or 5% casein in TBST instead of milk
Primary/secondary antibody washes are inadequate	Increase the wash numbers from 3 to 4. Increase the wash time from 10 to 15 minIncrease stringency of the wash buffer—increase NaCl and/or Tween 20 by 1.5–2-fold
Too much target protein was loaded	When a target band is too abundant on the membrane, washing off excess primary/secondary antibody may require longer washes and potentially more washes. Load 2–3-fold less protein. Increase the wash numbers from 3 to 4. Increase the wash time from 10 to 15 min
ECL is inadequate	Ensure the ECL solutions were prepared correctly and are not expired. Always use very fresh H_2O_2 for home-made ECL (<6 months since time of purchase)
Too much primary/secondary antibody was used	Decrease the antibody concentration by 2–3-fold. Decrease incubation time. Decrease incubation temperature. Overnight incubations at 4°C usually give more specific results
Low quality primary antibody	Preclear the antibody with a membrane containing lysates of cells with a knockout or knockdown for the target protein. See Section 4.2.5. Obtain a better commercially-available antibody. Decrease antibody concentration by 1.5–3-fold. Incubate at 4°C, instead of RT. Incubate for 1.5-fold shorter time. Dilute antibody in 5% BSA/TBST or 10% milk/TBST. Increase NaCl and/or Tween 20 concentration of TBST used for primary antibody solution and/or washes by 1.5–2-fold
The exposure is too high	At very high exposures, even the best antibodies will give off-target bands. Decrease exposure time by 2–3-fold
ECL was not drained well	Remove excess ECL on the membranes before and after they secured between two plastic films (transparencies)

4.3.5 Problem Type 5
Phospho-specific antibody detects nonphosphorylated protein

Potential Problems	Potential Solutions
Too much protein loaded per lane	Decrease protein amount loaded per lane by 2–3-fold. Loading more than 20 µg total protein per lane will result in nonspecific bands. Use protein quantification assays such as Bradford or BCA assay
Too much primary/secondary antibody was used	Decrease the antibody concentration by 2–3-fold. Decrease incubation time. Decrease incubation temperature. Overnight incubations at 4°C usually give more specific results

(Continued)

(Continued)	
Potential Problems	**Potential Solutions**
The antibody is not specific enough	Preclear the antibody with a membrane containing lysates of cells with a knockout or knockdown for the target protein. Phospho-specific antibodies can be absorbed onto a nonphosphorylated peptide column to remove nonspecific antibody molecules. Supplement the antibody solution with $10\ \mu g/mL$ nonphosphorylated peptide (for $1\ \mu g/mL$ antibody solution) during incubation with the target membrane. Obtain a better antibody

4.3.6 Problem Type 6
Bubbles were observed on the membrane

Potential Problems	**Potential Solutions**
Bubbles not removed properly during transfer sandwich assembly	Remove any potentially trapped air bubbles using a roller or a serological pipette to gently roll over the surface of both membrane and blotting papers
Blotting paper or membranes were not prewet long enough	Ensure that the blotting paper is prewet for ~ 15 min and the membranes for ~ 5 min
The transfer was done at very high voltage	Immerse tank in ice-water bath. Transfer at constant 100 V or constant 0.4A (for 12-gel tanks) for 1–2 h
Particles were trapped between gels and membranes	Rinse the gel glasses with water and gels with transfer buffer before starting transfer sandwich assembly. Prewet membranes in fresh, clean transfer buffer

4.3.7 Problem Type 7
Gels will not start running

Potential Problems	**Potential Solutions**
Running buffer was not prepared properly	Make fresh running buffer
Power supply does not work	Try a new power supply
Loose electrode contact	Tighten the electrodes
Loose running tank lid	Change the lid
Level of the running buffer in the inner or outer compartment is too high or too low	Adjust the running buffer levels

4.3.8 Problem Type 8
Transfer will not start

Potential Problems	**Potential Solutions**
Transfer buffer was not prepared properly	Make fresh transfer buffer and prechill before use
Inadequate power supply	Some power packs cannot handle resistance higher than a certain level. Large tanks, for example, cannot be run on small power packs

(Continued)

(Continued)	
Potential Problems	**Potential Solutions**
Loose electrode contact	Tighten the electrodes
Loose transfer tank lid	Change the lid
Not enough transfer buffer	Fill the transfer tank completely

4.3.9 Problem Type 9
"Smiling" effect during elecrophoresis

Potential Problems	**Potential Solutions**
Running buffer causes high resistance and therefore band migration is faster in the center	Decrease voltage. Ensure that the running buffer levels are adequate in the inner and outer cell. If the gel is running for prolonged time, immerse the tank in an ice-water bath. Make fresh running buffer
Too much or uneven protein loaded	Load the same amount of protein in each lane. If too much protein needs to be loaded, then run the gel very slowly
Low gel quality or expired precast gel	Mix the component of the gel properly. Avoid bubbles or particles in the gel
Low quality sample—genomic DNA in the sample	If the sample is too viscous, sonicate it to shear the genomic DNA. Benzonase can be used instead of sonication
Low quality sample—too much glycerol or detergent in the sample	Dilute 2–3-fold if the final detergent or glycerol concentration is too high in the sample

4.3.10 Problem Type 10
Miscellaneous

Potential Problems	**Potential Solution**
Bands with white lines in them	Load 3–5-fold less protein. Decrease primary and/or secondary antibody concentration by 2–3 fold. Dilute ECL solution by twofold
Black dots on the blot	Antibodies are reacting with the blocking solution. The blocking solution should be filtered. Exposure cassette needs to be cleaned. Exposure is too long. Reduce exposure time

CHAPTER 5

Tips and Tricks

A creative man is motivated by the desire to achieve, not by the desire to beat others.

—Ayn Rand

Don't wait for extraordinary opportunities. Seize common occasions and make them great!

—Orison Swett Marden

Come at once if convenient. If inconvenient, come all the same.

—Arthur Conan Doyle, Sherlock Holmes

#1: Choose wet transfer, if possible

Semidry transfer seems easier and requires less buffer volume, but the "prettiest" blots (i.e. consistent high-quality results) usually are obtained with wet transfer tanks. Moreover, wet transfer tanks have usually a higher capacity for simultaneous transfer of multiple gels and this is an important lab efficiency factor.

Western Blotting Guru. DOI: http://dx.doi.org/10.1016/B978-0-12-813537-2.00005-9

#2: Avoid overloading and viscosity

One of the best ways to get perfect bands is to load 20 μg or less of total protein per lane, if the antibody sensitivity and the antigen abundance permits. Also, try to make sure that the samples are not viscous. Genomic DNA should be sheared using Benzonase® or sonication to avoid viscosity. Excessive levels of chemicals such as urea and detergents should be diluted, if possible. Benzonase is very active and will work even in the presence of sodium dodecyl sulfate (SDS) and urea, as long as ethylenediaminetetraacetic acid (EDTA) is not present in the sample. If EDTA is present, add 7 mM of magnesium acetate per 1 mM of EDTA present in the sample to neutralize EDTA.

#3: Run and transfer slowly

It is tempting to run the gels and transfer them rapidly and save time, but high-quality runs and transfers are easier to obtain at constant low voltage/current, which then requires longer run/transfer times. For a BioRad Tetra system, 75 V for the whole run works best, however, if the sample quality is high enough (i.e. non-viscous, low detergent levels) and the wells are not overloaded, the run can be started and finished at 200V, without losing band quality. For a 12-gel transfer using BioRad's large TransBlot cassette, 0.4 A for 2 hours results in transfer of even very large proteins like DNAPK (469 kDa). If you are going to do the longruns and long transfers, the easiest way to maintain cold temperature and avoid problems during runs/transfers is either perform them in the cold room or simply immerse the tanks into ice-water. Cooling coils and refreezable ice-blocks are also available for some tank types.

#4: Premix gels Water, buffer, and SDS can be premixed and stored at room temperature, so that you only need to add acrylamide, ammonium persulfate (APS), and tetramethylethylenediamine (TEMED) on the day of gel casting. However, best quality is obtained when mixing all the components on the day of casting.

#5: Dilute the marker and mark after transfer

Prestained molecular weight markers can be diluted fivefold to make them last longer. Naturally, the bands will be fainter after transfer, but marking them with a blue ballpoint pen will make the marker last longer after multiple stripping rounds.

Recommended: Staedtler, 430-M.

#6: Fast strip

A rapid and easy stripping solution that doesn't involve malodorous beta-mercaptoethanol or heating up anything: 0.2 M NaOH + 2% SDS. Rinse the blots once with this solution and then add the solution again and incubate at room temperature for 10 minutes. Wash with TBST (Tris-Buffered Saline Tween 20) four times for 5 minutes. No need to reblock for most antibodies.

Note: Sodium hydroxide (NaOH) is highly corrosive. Use necessary personal protective equipment.

Note: Sodium dodecyl sulfate (SDS) is flammable, corrosive, irritant, and highly toxic. Use necessary personal protective equipment.

#7: Permanent cassette-film corner label

Instead of cutting or bending the edges of films before exposing them, place a phosphorescent tape to the corner of the cassette (1−2-mm-thick strip is enough). This way you can have multiple cassettes with permanent cassette-specific markings made with different sizes and shapes of the phosphorescent tape. This will make it easier to organize film/cassette combinations when handling more than one at once. If the tape is too bright for your developer settings, tape a Scotch tape over it to reduce the brightness.

Recommended: ProTapes Pro Glow Phosphorescent Vinyl Glow— available from Amazon.com.

#8: Smart elution from beads

When eluting immunoprecipitated proteins from antibodies conjugated to beads (e.g., anti-Flag-agarose, anti-myc-agarose, and anti-HA-agarose beads), it is best to keep the heavy chain of the antibody on the beads. To do this, after the last immunoprecipitation (IP) wash, add 3% SDS (three volumes of the beads, i.e., 75 μL of 3% SDS to 25 μL beads) and shake at 55°C for 15 minutes at 1200 rpm. Spin the beads (1000 × g, 30 seconds, 25°C) and take the supernatant. Add 5X sodium dodecyl sulfate-polyacrylamide gel electrophoresis (SDS−PAGE) sample buffer to this supernatant to 1 × final concentration. Heat at 70°C for 10 minutes and load. This way, if the bait or the prey in the IP is around 50 kDa, you will not get interference from the antibody heavy chain, which is also ∼50 kDa, even if the secondary antibody is the same host (e.g., anti-mouse-horse radish peroxidase (HRP) secondary used for an anti-FLAG® IP with the mouse monoclonal M2 antibody). The antibodies can be also conjugated/crosslinked to protein A/G-sepharose beads in-house (see Appendix K).

#9: Different kind of secondary antibody

Alternatively, protein A/G-HRP conjugate or a secondary that detects only native form of antibodies (e.g., TrueBlot) can be used as a secondary antibody, in order to avoid detecting heavy chains on the blot. However, these can turn out to be tricky to optimize, depending on the case and the most robust way to avoid heavy chain detection is covalent crosslinking (conjugation) of the antibodies to the beads (see Appendix K).

#10: Boil the membrane

For some antibodies, such as certain anti-ubiquitin antibodies, briefly boiling the membrane after the transfer (in TBST, in microwave, 5 seconds) can increase sensitivity to the antigen.

#11: Properly reuse primary antibodies

Most primary antibody dilutions can be reused 2−3 times; some can be reused even more than that. The best way to store a used antibody dilution is to supplement it with 0.05% sodium azide (NaN$_3$) and keep it at 4°C. Antibody dilutions can be stored in this manner for 1−2 months, depending on the antibody stability. For long-term storage (2−6 months) it's recommended to freeze the dilutions at −20°C. Every antibody is different, so whether an antibody survives such freeze−thaw (after being diluted) or not has to be empirically determined. It is best not to reuse secondary antibodies.

Note: Sodium azide is highly toxic and hazardous to the aquatic environment. Use necessary personal protective equipment.

#12: Drying the gel glasses

The best way to dry glasses is to wash them one day in advance and let them dry overnight. But, if you're in a hurry, an aspirator filter can be attached to the "air" valve on the lab bench to make the pressurized air come out in a focused beam and to filter out potential dirt particles that may come out. Such "clean" air stream can then be used to dry the glasses relatively quickly, albeit, noisily. Also, glasses can be given a rinse with ethanol from a squeeze bottle and placed on a tissue paper in a fume hood, to dry them out even faster.

#13: Casting gels and sealing with ethanol—squeeze bottles

Ethanol can be poured into a squeeze bottle to make the sealing of resolving gels faster, which is especially useful when casting multiple gels. Ethanol also dries out faster than water, so you can pour the stacker sooner after the resolving gel polymerizes and the ethanol is decanted/aspirated from the top of the gel glasses. Also, large volumes of gel mixtures can be prepared in a squeeze bottle to make the casting of the resolving and stacker easier and faster.

#14: Storing the membranes

Once you are done exposing the membranes, they can be washed with TBST (5 minutes), rinsed briefly with distilled water, dried and stored at room temperature for at least 6 months!

#15: Storing the gels (short term)

The best way to store gels is to interspace them with wet tissue paper, wrap with saran plastic wrap (without removing the combs) and keep at 4°C for less than a week. However, this can be tedious and generates plastic waste. For shorter times (2−3 days), the gels can be simply put into a box (without removing the combs), covered with distilled water and stored at 4°C.

#16: Multiple antibodies trick

Multiple antibodies can be combined in the same incubation box for some highly-specific antibodies that give only a single-band result. We have successfully tried up to 6 antibodies in a single box. This is best done for proteins of different molecular weights and when you're not looking for mobility shifts. Ensure that you have a positive and a negative control. An extra primary antibody wash can be helpful to reduce overall background.

#17: Primary + secondary incubation trick

For very strong antibodies with abundant antigens (e.g., tubulin, actin, GAPDH) primary and secondary antibodies can be combined and incubated together at room temperature for only 30 minutes. Then proceed with 3×10 minutes washes with TBST and develop the blots.

#18: Use a pencil to label membranes to be used for fluorescent detection

If you mark the membranes with a ballpoint pen, you will leave traces that will interfere with systems like LICOR and leave smears. However, a regular pencil will overcome this problem, as pencil marks are undetectable to such systems.

#19: Amplification by sandwich

For very weak signals, DIG (digoxigenin)/biotinylated-modified antimouse/rabbit-secondary antibody can be sandwiched between the primary and the anti-DIG-HRP (or streptavidin-HRP) tertiary antibody to amplify the signal. However, consider that this trick extends the procedure time and the background levels. Moreover, first, consider employing "sensitive films" (intended to use for autoradiography) and/or commercial enhanced chemiluminescence (ECL) solutions such as Femto™ and Pico™ (GE Healthcare Life Sciences/Amersham) to circumvent a low sensitivity/low signal levels problem you may be having.

#20: SDS power

To increase the transfer efficiency of large proteins, add SDS to 0.1% into the transfer buffer. However, this will decrease binding of the proteins to nitrocellulose membranes, so use polyvinylidene fluoride (PVDF) membranes for large proteins. Presence of SDS will increase the amount of heat generated, so make sure to place the tank into ice-water bath and monitor the temperature to avoid overheating and bubble formation.

#21: The pH is alright

The pH of these buffers will be dictated by the Tris and Glycine present in the solution at it should be around 8.8 for running and 8.3 for transfer buffer after all the ingredients are fully dissolved. Adjusting pH will change the salt concentration and therefore conductivity of the buffers, so if the pH is incorrect, do not adjust the pH, but re-prepare the buffers.

#21: Tween tweak

For weak antibodies, the Tween 20 concentration in the TBST wash steps can be reduced from 0.1% to 0.03% or 0.01%. For dirtier antibodies, 0.1% can be increased to 0.2%. It is easier to handle Tween 20 using a syringe or preparing a 20% stock, which is much less viscous than 100% and therefore the pipetting errors will be much smaller.

#22: EDTA preservative

Adding EDTA to 1 mM final concentration in the 10X TBST stock will inhibit potential contamination growth and increase the shelf life of the stock. Avoid using NaN_3 (sodium azide) in TBST solutions or solutions in which HRP-conjugated secondary antibodies will be diluted, as NaN_3 is a potent HRP inhibitor and will result in no ECL signal.

#23: Storing the gels (long term)

If the resolving gel only is poured and the gels are immersed in running buffer, the gels can be stored for about $1-2$ months. Keep in mind that immersing gels in liquids may result in glass plate separation, so handle them carefully when removing from the storage.

#24: Of spacers and dyes

To avoid irregular band shape and migration, always load samples with comparable density side-by-side. Leave a spacer lane between a molecular weight marker and the first sample and load a $1\times$ dye there. That way the usually different density of the marker won't affect the band shape and migration of the samples in the first sample lane.

#25: Non-phosphopeptides

To increase the specificity of a phospho-specific antibody, non-phosphopeptides can be directly added to the solution of the primary antibody during incubation with the blot. Unless the phospho-specific antibody is a monoclonal or has been affinity purified against a peptide column, this trick many times becomes essential. Synthetic peptides can be relatively expensive, therefore, alternatively Glutathione S-Transferase (GST)-fusion of the peptide can be expressed in bacteria and stocked. Such short peptide fusions of GST usually express at very high levels in bacteria and therefore, even from 0.5 to 1 L of culture a long-lasting stock of a GST-non-phosphopeptide fusion can be prepared.

Recommended: pGEX6P2 vector and BL21 *Escherichia coli* strain.

#26: Speed up the cast

To speed up the gel casting, after adding APS and TEMED, and pouring the gel solutions into the glasses, the gel stands can be moved to a 37°C dry incubator to halve the time required to finish the polyacrylamide gel matrix polymerization. Keep in mind that longer incubation at 37°C will dry out the gels, therefore, limit the time to 30 minutes. In addition, increasing APS and TEMED concentrations by 30% will also shorten the time required to finish the polymerization. When this trick is used: If casting multiple gels, divide the gel premix into 2−3 batches to prevent premature hardening of the gels before water pouring or comb insertion step.

#27: Quick lysis

Cells can be lyzed in 12-well plates or 6-well plates, using SDS−PAGE-ready 1X SLB buffer. This will result in a viscous lysate that is ready to be heated, spun and loaded into the SDS−PAGE gels.

To remove the viscosity, either briefly sonicate the lysates or add Benzonase® (see Appendix I for detailed protocol).

#28: Batch scan

Many people cut up their immunoblotting exposures/films (X-ray films) and scan only the relevant exposures. However, this can result in loss of valuable data, unappreciated at the time of the scan, but potentially meaningful at a later stage, as projects unravel. Therefore, we routinely scan our films using an **automatic document feeder** available in most modern scanners, in order to simplify and expedite the scanning process and be able to effortlessly save all, rather than "currently relevant" exposures. This also makes comparison of various exposures easier, as it can be done on-screen after scanning all the films.

Recommended: Canon PIXMA MX452 printer/scanner/copier.

#29: How to load more

If the amount of samples to be loaded is more than the volume of the well, you can create a double-well by aspirating the wall between two wells. This is particularly useful for applications where the band of interest is going to be cut out of the gel (e.g. for processing through mass spectrometry).

#30: How not to tear the gel

Remove one of the gel glasses. Place a pre-wet blotting paper (or membrane) the size of the gel on top of the gel. Invert this sandwich and gold it glass facing up. Dislodge the gel from the second glass, using forceps or a plastic wedge tool and let the gravity to aid the gel-glass separation process.

CHAPTER 6

Special Cases

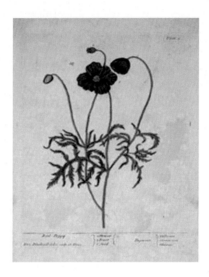

The highest reward for a man's toil is not what he gets for it but what he becomes by it.

—*John Ruskin*

It's not the things that are precious. It's the idea behind them. What they can do.
—*Alec Sadler*

You must have long range goals to keep you from being frustrated by short range failures.

—*Charles Noble*

Hi. We're not weird. We just seem like we are.

—*Rex Van de Kamp*

6.1 QUANTITATIVE WESTERN BLOTTING

Inherently, due to the usage of the secondary antibodies and the amplification steps in the Western blotting, the method is nonlinear and therefore semiquantitative. LI-COR and other companies provide fluorescently conjugated secondary antibodies and fluorescent flatbed

Western Blotting Guru. DOI: http://dx.doi.org/10.1016/B978-0-12-813537-2.00006-0

scanners that can be substituted for enhanced chemiluminescence (ECL) / film-based detection of the signal. Such experimental design results in lack of amplification of the signal due to the linear nature of the quantified fluorescent signal. Notably, fluorescent secondary antibodies also allow multiplexing of Western blotting, which can be vital for some experiments. For example, total and phospho-specific signals can be visualized simultaneously, saving time and giving precision in analysis, which otherwise could be lost due to stripping/reprobing or rerunning the samples. Moreover, analysis of cleavage fragments using N-terminal and C-terminal antibodies can be performed, which otherwise could be difficult to analyze due to potential close proximity of the fragment sizes.

6.2 OVERLAY ASSAYS

For certain applications, the usage of antibodies during immunoblotting can be substituted with digoxigenin (DIG)-labeled proteins that have high affinities for their targets. In other words, the protein probe acts as a primary antibody and anti-DIG-HRP is used as a secondary antibody. Such variations of Western blotting are called "overlay assays". A popular example of overlay assays is 14-3-3 overlay assay, where yeast 14-3-3 isoforms are purified and conjugated to DIG to be used as a primary antibody to probe for Ser/Thr phosphorylation sites that form binding sites for these phosphorylation sensor proteins (see Mackintosh, 2004; Johnson *et al.*, 2010).

14-3-3 overlay assay protocol (adapted from Moorhead *et al.*, 1996):

1. **Block:** 1 hour in TBST containing 5% milk and 0.5 M NaCl.
2. **Probe:** with 5 µg/mL total DIG-BMH1 and DIG-BMH2 (yeast 14-3-3 isoforms, expressed in *E. coli*), in TBST containing 1 mg/mL BSA and 0.5 M NaCl for 16 hours at 4°C.
3. **Wash:** 6 times for 5 minutes with TBST containing 0.5 M NaCl
4. **Probe:** with a 1:5000 dilution of anti-DIG-HRP (anti-digoxigenin antibody conjugated to horse radish peroxidase) antibody in TBST containing 5% milk and 0.5 M NaCl, for 1 hour at RT.
5. **Wash:** 5 times for 5 minutes with TBST.

6.3 PHOSPHO-SPECIFIC ANTIBODIES

Antibodies can be raised against virtually anything that is immunogenic enough in a given host animal. One of the most popular applications in molecular biology is raising antibodies against phosphorylated proteins. In fact, phospho-specific polyclonal and especially monoclonal antibodies are vital tools for studying cell signaling pathways. Phospho-specific antibodies are generated using a short peptide that is chemically synthesized to contain a phosphorylated residue and is used as an immunogen upon conjugation to an adjuvant/carrier such as Keyhole Limpet Hemocyanin (KLH).

Phosphorylation of only four amino acid types is known so far— Ser, Thr, Tyr, and His. Ser/Thr phosphorylation is by far the most widely-studied type and most of the kinases in the human kinome are Ser/Thr kinases (www.kinase.com). Anti-phospho-Ser/Thr antibodies that are designed to detect any phosphorylated Ser/Thr residue exist and can be used in some applications. Tyr phosphorylation is less studied, however, the available anti-phospho-Tyr antibodies usually have higher affinity than their Ser/Thr counterparts. Histidine phosphorylation is the least studied type and phospho-specific antibodies for p-His have recently been developed (see Fuhs *et al.*, 2015).

One of the key factors is the specificity of phospho-specific antibodies. During monoclonal antibody generation, this can be achieved by screening for a clone that detects a phosphorylated, but not a nonphosphorylated peptide, using Dot blots (see Section 6.6 for Dot Blot). However, monoclonal antibody generation is an expensive and time-consuming process; thus, many laboratories that need phospho-specific antibodies raise polyclonal antibodies. For polyclonal phospho-specific antibodies they key to phospho-specificity is rigorous purification of the sera over phospho-peptide and/or nonphosphopeptide columns. Many commercial companies offer comprehensive services for antigen design, mouse or rabbit immunization and sera purification steps of the polyclonal phospho-specific antibody generation process.

The factors to consider while generating antibodies against phosphosite(s) include the number of close-by phosphorylated residues. Although monoclonal antibody generation can result in clones that can detect multiple phosphorylation sites in a single peptide, achieving the same result with polyclonal approach can be difficult.

6.4 PHOS-TAG

Studying phosphorylation of proteins using phospho-specific antibodies can be sometimes complicates by the presence of multiple close-by phosphorylated residues, raising reliable antibodies against which can be

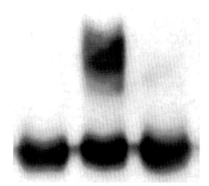

*Figure 6.1 **Robust phosphorylation detection using Phos-tag.** Cells were transfected with Flag-tagged protein of interest and treated 48 h later. Lane 1: untreated cell lysate. Lane 2: lysate of cells treated with forskolin for 1 h. Lane 3: as in Lane 2, except in the presence of PKA inhibitor H-89. 50 μM Phos-tag was used in the resolving gel. Membranes were probed for Flag.*

tricky (i.e., more than 1 phosphosite in a 5-amino acids sequence). A method involving usage of a chemical called Phos-Tag, a phosphate-interacting compound that can be supplemented into the polyacrylamide gels, allows analysis of protein phosphorylation by immunoblotting for the total target protein and observing phosphorylation-dependent mobility shifts (Figure 6.1.). This phosphorylation-dependent nature of the mobility shift can be tested using *in vitro* phosphatase assays (that would remove the phosphates and collapse the shifts) and/or Ser/Ala (or Tyr/Phe) mutant analysis, to study individual sites contributing to the mobility shifts. The disadvantage of this method is that the target protein should usually be below 75 kDa to work robustly under canonical conditions of the method. However, upon careful optimization, larger proteins can be analyzed used this method too.

6.5 NONREDUCING PAGE

For some protein oligomers, the 0.1% Sodium Dodecyl Sulfate (SDS) present in the running buffer and SDS−PAGE gels is not enough to denature the oligomers into monomers, but is enough to coat them and give them a negative charge that will facilitate their migration in the gel. This allows running samples on regular SDS−PAGE gels and a rapid, simple protocol for analyzing oligomerization of proteins (proper native gels involve a more complicated and slower protocol). The difference between nonreducing PAGE and regular SDS−PAGE is the lack of usage of reducing agents (e.g. beta-mercaptoethanol (BME), dithiothreitol (DTT)) and lack of heating of the samples prior to loading. Keep in mind that certain antibodies require their target epitopes to be exposed by denaturation of the protein structure.

6.6 DOT BLOTS

Dot blots are a rapid alternative to canonical immunoblotting, but lack the separation by size dimension that SDS−PAGE gives. Dot blots can be employed to test antibody reactivity with phospho-peptides. The quality of an antibody to be used on dot blots for probing lysates has to be very high, in order to produce meaningful results.

Dot Blot Protocol:

1. Draw a grid, using a pencil, on a sheet of nitrocellulose membrane.
2. Place 2−3 μL of sample onto the dry nitrocellulose membrane. Use different concentrations of the sample solution and apply the same volume to keep the size of the circle the same.
3. Air dry the membranes.
4. Block the membranes in 10% milk in TBST for 30 minute−1 hour.
5. Continue as for canonical immunoblotting.

CHAPTER 7

Data Analysis, Storage, Retrieval

Perfectly sound analysis, but I was hoping you'd go deeper.
—Sherlock Holmes (Sherlock, 2010—)

If you want to succeed you should strike out on new paths rather than travel the worn paths of accepted success.
—John D. Rockefeller

Victory belongs to the most persevering.

—Napoleon

One of the most important steps in immunoblotting is proper data analysis, storage and retrieval system. Many times, data, exposures, or parts of the blots deemed uninformative become quite important as projects unfold and new relevant information becomes available. In this chapter, we focus on series of practices that form a versatile system for data analysis, storage and retrieval.

Western Blotting Guru. DOI: http://dx.doi.org/10.1016/B978-0-12-813537-2.00007-2

#1: Scan and save all exposures (also mentioned in Chapter 5, Tips and Tricks, #29).

Many people cut up their immunoblotting exposures/films (X-ray films) and scan only the relevant exposures. However, this can result in loss of valuable data, unappreciated at the time of the scan, but potentially meaningful at a later stage, as projects unravel. Therefore, we routinely scan our films using an automatic document feeder available in most modern scanners, in order to simplify and expedite the scanning process and be able to effortlessly save all, rather than "currently relevant" exposures. This also makes comparison of various exposures easier, as it can be done on-screen, after scanning all the films.

We recommend scanning films at 300 dpi resolution. This is sufficient for most, if not all, journal requirements for immunoblotting images. Scanning films at higher resolution results in much larger images, which are harder to handle in image processing programs (e.g., Photoshop, Fireworks). Moreover, it becomes more expensive to store large numbers of images.

It is useful to save the original films for potential future rescan needs, in case, either higher resolution, dust/speck removal or better contrast is needed.

Scan and store all the films in a dated format to fit the Excel organizer system described below.

#2: Setup an Excel organizer file to make experimental information and data panels searchable

If every experiment is entered into an Excel table with a unique dated identifier (see identifier "161230b"—2016-12-30b in Figure 7.1), one can easily search an Excel-based database of each immunoblotting session and retrieve relevant information such as loading order, sample type, antibodies used for probing and reprobing, and volumes loaded.

	10% (29:1)/1.5mm	Jan_17	#			Probe	Load
1	161230b IP-HA- mTLR2 RGC-5 MF LPS tc	vec (MF-10mM-0, 0.5, 1, 2h, LPS-15min, 30min); mTLR2-HA (")	12	Lys	1	p-ACC + p-AMPK + p-S6	20µl
2					2		"
3					3		"
4					4		"

Figure 7.1 A sample Excel organizer table used for storing experimental data.

If image files containing data panels for this experiment are named with the same unique identifier, the experimental information stored in this Excel file can be matched to this results file and such information will become instantly searchable.

This "unique identifier" approach combines the versatility of storing and organizing detailed experimental information in an Excel table with the ability to instantly find the relevant image data file using the identifier, and vice versa.

#3: Have a template panel of labels saved in image-editing software format

When working on a project, many times the antibodies and experimental setups employed across several experiments are recurrent. Thus, time-saving labeling template files can be used to simply the data paneling/assembly sessions (Figure 7.2).

#3: Where appropriate, digitize/quantify the bands and normalize to internal controls

Although a reliable band signal quantification can be obtained using fluorescently-labeled antibodies employing fluorescent scanners and their accompanying software, CCD camera imaging devices can be employed to perform semiquantification of the western blotting signals generated using ECL (using accompanying software or third-party applications). If X-ray films were used for ECL signal visualization, following scanning of the films, semiquantification of the bands can be done using several programs, including Image J, which is a free software package provided by the NIH (https://imagej.nih.gov/ij/).

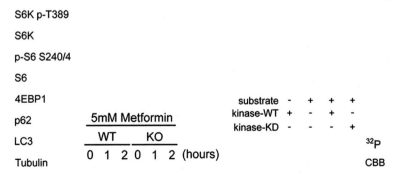

Figure 7.2 A sample data file labeling template.

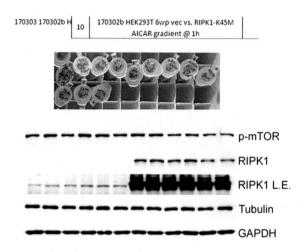

Figure 7.3 A sample labeling data file with the tube label photograph.

Once the band signal is quantified, normalization of phospho-signal to total or total target protein signal to an internal loading control, such as actin or tubulin should be done.

#4: Link samples to the data using photos of the tubes

It is often necessary to return and rerun certain samples. Therefore, to ensure that the correct sample sets are being rerun, an easy way to link a sample set to a data panel is to place a photograph of the tubes into the data file. This will also provide an easy method for tracking the order and orientation of the loaded samples (Figure 7.3).

#5: Take notes inside the data files

Although extensive notes can and should be taken elsewhere, it is often practical to write down deductions, observations, and conclusions right inside the data files. This will ensure that the precious ideas are not lost and will make it easier to return and analyze to old data files. Moreover, future to-do tasks can also be written inside the data files, to simplify follow ups.

#6: Save data files with a lot of information in the filename

Consider the filename of the data files as a metadata for your experiment. Information such as date, cell or tissue type, drugs or treatment used, length of treatment, and other unique information should be placed into the filename. Such naming of the files will allow simplified and rapid retrieval of the data files upon query for desired parameters, when the files are searched for on the users computer.

Several examples of filenames for western blotting data files:

160131 siCyld in RGC-5 cells
150128 Glucose starvation timecourse in cIAP1-2 dKO MEFs
Placing date in yy-mm-dd format allows sorting by filename sort automatically by the experiment date.

Buffers and Solutions

40% Acrylamide Stock (37.5:1)

Acrylamide	195 g
Bis-acrylamide	5.2 g
H_2O	to 500 mL

Preferably, filter through 0.45 μm filter.

Store in the dark at 4°C (wrap the bottle in foil).

Acrylamide is neurotoxic and should be handled with care.

1.5 M Tris−Cl (pH 8.8)

Tris base	36.3 g
H_2O	To 200 mL

Adjust pH to 8.8 with HCl. Preferably, filter through 0.45 μm filter. Store at room temperature (RT).

0.5 M Tris−Cl (pH 6.8)

Tris base	6.06 g
H_2O	To 100 mL

Adjust pH to 6.8 with HCl. Preferably, filter through 0.45 μm filter. Store at RT.

10% Sodium Dodecyl Sulfate (SDS)

SDS	10 g
H_2O	To 100 mL. Preferably, filter through 0.45 μm filter. Store at RT

10% Ammonium Persulfate

Ammonium persulfate	1 g
H₂O	10 mL

Preferably, filter through 0.45 μm filter. Store at −20°C in single-use aliquots.

5× SDS-polyacrylamide gel electrophoresis (PAGE) sample dye (200 mL)

(315 mM Tris–HCl, pH 6.8, 10% SDS, 5% beta-mercaptoethanol, 0.01% bromophenol blue (BPB), 50% glycerol)

1. Add **63 mL of 1M Tris–HCl pH 6.8** into a 250-mL flask with a magnetic stirrer. Start the stirrer.
2. Add **100-mL glycerol** using a needle-free syringe. Prewarmed glycerol to 37°C to make it easier to handle.
3. Add **20-g SDS powder**. Stir for 5 min.
4. Place the flask into a 55°C water bath for 10 min.
5. Stir for another 15−20 min at RT.
6. Add **10-mL beta-mercaptoethanol** (in the fume hood).
7. Add **2 mL 1% BPB** to desired darkness of the dye (dilute a 50-μL aliquot to 1× to assess the darkness).
8. Add H₂O to 200 mL final. Stir for another 15−20 min at RT.
9. Make 1 mL aliquots. Freeze at −20°C. Thaw by incubating at 90°C for 30 s. Avoid refreezing more than 3−4 times.

Native Lysis Buffer

Filter the buffer (0.45 μm filter bottle) and store at 4°C for up to 2 months.

Store at −20°C (manual defrost freezers) for up to 6 months.

25 mM Hepes, pH 7.4, 1% Triton X-100 (or 0.2% NP-40 or 0.3% CHAPS), 120 mM NaCl, 0.27 M sucrose, 5 mM EDTA, 5 mM EGTA, 50 mM NaF, 10 mM b-glycerophosphate, 5 mM sodium pyrophosphate, 2 mM benzamidine (fresh), 0.1% BME (fresh) (or 1 mM DTT), 1 mM PMSF (fresh), 2× Roche's Complete protease inhibitor cocktail.

Final	Stock	Take
40 mM Tris–HCl, pH 7.5 or	1 M	40 mL or
25 mM HEPES-KOH, pH 7.5	1 M	25 mL
0.3% Chaps or	Powder	3 g or
1% Triton X-100 or	100%	10 mL or
0.2% NP-40	100%	2 mL
120 mM NaCl	4 M	30 mL
0.27 M Sucrose	Powder	92.4 g
5 mM EDTA	0.5 M	10 mL
5 mM EGTA	0.5 M	10 mL
50 mM NaF	Powder	2.1 g
10 mM beta-glycerophosphate	Powder	3.14 g
5 mM sodium pyrophosphate	Powder	1.11 g of dibasic, 1.33 g of decahydrate
1 mM Na_3VO_4	200 mM	**1:100 (fresh)**
Complete protease inhibitor cocktail	$100 \times$ [a]	**1:100 (fresh)**
Benzamidine	1 M	**1:1000 (fresh)**
PMSF	0.1 M	**1:100 (fresh)**
BME	14.3 M	**1:1000 (fresh)**
DTT	1 M	**1:1000 (fresh)**

[a]*Prepare $100 \times$ stock by dissolving 1 tablet meant for 50 mL in 500 μL of water.*

SDS−PAGE Gel Tables

	1 gel	1 gel	1 gel	1 gel	1 gel
	7 mL	7 mL	7 mL	7 mL	7 mL
	6%	**8%**	**10%**	**12%**	**15%**
H₂O	4.05 mL	3.7 mL	3.5 mL	3 mL	2.475 mL
1.5M Tris−HCl, pH 8.6	1.8 mL	1.8 mL	1.75 mL	1.8 mL	1.8 mL
40% Acrylamide	1.05 mL	1.4 mL	1.75 mL	2.1 mL	2.625 mL
10% SDS	70 μL	70 μL	70 μL	70 μL	70 μL
10% APS	35 μL	35 μL	35 μL	35 μL	35 μL
TEMED	3.5 μL	3.5 μL	3.5 μL	3.5 μL	3.5 μL

	1 gel	2 gels	4 gels	6 gels	8 gels	10 gels	12 gels
	10%	**10%**	**10%**	**10%**	**10%**	**10%**	**10%**
H₂O	3.5 mL	7 mL	14 mL	21 mL	28 mL	35 mL	42 mL
1.5M Tris−HCl, pH 8.6	1.75 mL	3.5 mL	7 mL	11 mL	14 mL	18 mL	21 mL
40% acrylamide	1.75 mL	3.5 mL	7.0 mL	10.5 mL	14 mL	17.5 mL	21 mL
10% SDS	70 μL	140 μL	280 μL	420 μL	560 μL	700 μL	840 μL
10% APS	35 μL	70 μL	140 μL	210 μL	280 μL	350 μL	420 μL
TEMED	3.5 μL	7 μL	14 μL	21 μL	28 μL	35 μL	42 μL

	1 gel	2 gels	4 gels	6 gels	8 gels
	Stacker	**Stacker**	**Stacker**	**Stacker**	**Stacker**
H₂O	2.1 mL	4.2 mL	8.4 mL	12.6 mL	16.8 mL
1M Tris−HCl, pH 6.8	0.37 mL	0.74 mL	1.48 mL	2.23 mL	2.98 mL
40% Acrylamide	0.28 mL	0.57 mL	1.14 mL	1.7 mL	2.27 mL
10% SDS	30 μL	60 μL	120 μL	180 μL	240 μL
10% APS	26 μL	52 μL	104 μL	156 μL	208 μL
TEMED	3 μL	6 μL	12 μL	18 μL	24 μL

SDS—PAGE Protocol

Materials:

- Deionized/distilled water
- 40% Acrylamide stock (37.5:1)
- 10% ammonium persulfate (APS)
- TEMED
- 0.5 M Tris—HCl, pH 6.8
- 1.5 M Tris—HCl, pH 8.6
- **1× Running Buffer** (0.025 M Tris, 0.192 M glycine, 0.1% SDS, pH8.3). Recipe for 10× 1 L stock: 30 g Tris, 144 g glycine, 10 g SDS. Prepare 10× stock and store at room temperature up to 1 month.
- 1× Sample buffer (see Appendix A for 5× SDS—PAGE sample buffer recipe)
- Molecular weight marker
- Heat block

Procedure:

1. Wash the gel glasses and air dry until completely dry.
2. Assemble the glasses for gel pouring.
3. Prepare the **resolving gel** mix (see Appendix B) and cast it, leaving enough space for the stacking gel (about 5 mm from the tip of the comb is a long enough staking gel length).
4. After casting the resolving gel immediately and very slowly pour deionized water onto the gel, using a squeeze bottle or a serological pipette.
5. Let the gel to polymerize for 30 minutes at room temperature.
6. Aspirate/decant the water at the top of the separating gel.
7. Prepare the **stacking gel** mix (see Appendix B) and cast it.
8. Insert the combs immediately and slowly.
9. Let the gel to polymerize for 45 minutes—1 hour at 25°C.

10. Carefully pull the combs out and wash away the polyacrylamide pieces with distilled water.
11. Assemble the gels into tank.
12. Pour $1\times$ running buffer. The wells can be aspirated to clear the remaining pieces of acrylamide, if any.
13. Heat the samples at $70°C$ for 10 minutes or $95°C$ for 5 minutes. For most phospho-specific antibodies 20 µg of lysate is enough.
14. Invert the tubes 2−3 times and spin: $16,000\times g$, 1 min, $25°C$.
15. Load equal volumes in all wells. Fill empty wells with $1\times$ sample buffer and dilute the molecular weight marker to keep the volume constant across wells.
16. Run: **70 V** while the dye front is in stacker and then increase to **150 V** for resolving, until the dye front is just about to exit the gel.

Wet Transfer and Immunoblotting Protocol

Materials:

- **Blocking buffer:** 5–10% milk in TBST (prepare fresh on the day of usage).
- **Antibody buffer:** 5% BSA in TBST + 0.05% NaN_3. Prepare, filter through 0.45 μm filter, and store at 4°C. Avoid using NaN_3 in TBST solutions or solutions in which horse radish peroxidase (HRP)-conjugated secondary antibodies will be diluted, as NaN_3 is a potent HRP inhibitor and will result in no enhanced chemiluminescence signal.
- **Ponceau S staining solution:** 0.1% Ponceau S in 5% acetic acid.
- **1× Transfer buffer:** 48 mM Tris, 39 mM glycine, 20% ethanol (store a 10× stock at room temperature (RT). Dilute to 1× and add ethanol to 20% final on the day of usage). Recipe for 1 L 10× stock: 58.2 g Tris, 29.3 g glycine, mqH_2O to 1 L.
- **1× TBST (Tris buffered saline with Tween 20):** 50 mM Tris–HCl pH7.5, 150 mM NaCl, 0.1% Tween 20 (store a 10× stock at 4°C). Adding 1 mM EDTA to the 10× stock will prolong its shelf life. Dilute to 1× and store at RT. Prepare 1× volumes that will be used up in less than 2 weeks, to avoid contamination build up. Recipe for 1 L 10× stock: 87.7 g NaCl, 60.6 g Tris, 10 mL Tween 20. Adjust pH to 7.5 with 12 M HCl. Add mqH_2O to 1 L. Alternatively, mix 250 mL of 2 M Tris–HCl pH7.5, 375 mL of 4 M NaCl, 10 mL Tween 20 and add mqH_2O to 1 L.

Procedure:

1. Soak nitrocellulose membrane (1–3 minutes), 3 MM Whatman papers (15–20 minutes) and sponges (15–20 minutes) in **1× transfer buffer**.
2. The sandwich should be: (black side) sponge–paper–gel–membrane–paper–sponge (red/white side).

3. Mark the number of the gel using a ball pen on a region of the membrane that doesn't touch the gel.
4. Get rid of bubbles (a short serological pipette or a pen can be used).
5. Fill the tank with pre-chilled **1× transfer buffer**.
6. Transfer at 400 mA for 1.5−2 hours in ice/water bath.
7. Stain the blots in **Ponceau S staining solution** for 1 minute. Collect the staining solution for re-use (the solution can be re-used until it no longer gives proper staining results).
8. Destain the blots with mqH_2O.
9. Mark the blots with (1) molecular weight marker; (2) gel number; (3) date. Cut into slices and mark "T" ("top") or B ("bottom"), if necessary.
10. Incubate the blots for 30 minutes−1 hour in blocking solution.
11. Rinse the blots in **1× TBST**.
12. Incubate: Primary antibody in **antibody buffer** for 16 hours at 4°C on a side-to-side rocker.
13. Wash: Three times 5−10 minutes each with **1× TBST**.
14. Incubate: Secondary antibody (in **1× TBST**) for 1 hour at RT.
15. Wash: Three times 5−10 minutes each with **1× TBST**.
16. Combine solution 1 and solution 2 of the ECL reagent.
17. Add the ECL reagent to the blots.
18. Shake at RT for 2 minutes.
19. Assemble the exposure cassette.
20. Expose for 3 seconds, 10 seconds, 20 seconds, 30 seconds, 1 minute, 2 minutes, 10 minutes.

Notes:

- If using PVDF membrane, first soak it in methanol for 3−5 minutes, then in 1× transfer buffer.
- Ethanol or methanol can be used in the 1× transfer buffer.
- Decreasing ethanol/methanol from 20% to 10% would decrease the protein-membrane binding efficiency.

Home-Made Enhanced ChemiLuminescence (ECL) Detection

Materials:

Solution 1 (for 50 mL)

500 μL 250 mM Luminol (Sigma cat# A8511) (0.44 g/10 mL DMSO)

220 μL 90 mM *p*-coumaric acid (Sigma cat# C9008) (0.15 g/10 mL DMSO)

5 mL 1 M Tris−HCl, pH 8.6-8.8

H_2O to 50 mL

Solution 2 (for 50 mL)

30 μL 30% H_2O_2

5 mL 1 M Tris−HCl, pH 8.6-8.8

H_2O to 50 mL

Final concentrations:

2.5 mM Luminol, 0.4 mM *p*-coumaric acid, 100 mM Tris−HCl, pH 8.6-8.8, 0.018% H_2O_2

Procedure:

1. Combine 5 mL of solution and 5 mL of solution and shake.
2. Place up to four blots into a box with the combined ECL solution.
3. Shake at RT for 2 minutes.
4. Drain excess ECL solution and place the blots between two sheets of transparent plastic placed into an exposure cassette and fixed in place with small pieces of tape from the top two corners (common

transparent file folders work well, but pick the most transparent version of the folders).

5. Remove excess ECL solution and bubbles by passing a folded tissue paper over the upper plastic layer.
6. Place small pieces of tape to the bottom two corners of the file folder.
7. If different strength antibodies are used, exposing for 1, 10, 20, 30 seconds, 1, 2, 10 minutes is sufficient for most wanted levels of signal strengths.

Notes:

Since ECL reaction is not stabilized, unlike some commercial ECL kits, the procedure should be done as rapidly as possible to avoid loss of signal. Expose within 10−15 minutes after the films are removed from the ECL incubation solution.

- Usually the signal is lost after 25−30 minutes. Thus, exposures longer than that are usually meaningless.
- Prepare single-use luminol and *p*-coumaric acid aliquots and store in a manual defrost (i.e., not autodefrost) −20°C freezer.
- Mix Tris−HCl, luminol, and *p*-coumaric acid first, then add water.
- Make sure that the H_2O_2 is fresh! Buy fresh small aliquots and discard anything that is more than 6-month old.
- If the lowest exposure is still too dark, put two films on top of each other and expose like that.

Stripping Protocols

Long stripping protocol

Materials:

- 14.2 M beta-mercaptoethanol (BME)
- Basal stripping buffer (2% SDS, 62.5 mM Tris−HCl, pH 6.8).

Procedure:

1. Wash the membranes in TBST for 5 minutes, on a shaker.
2. Add 300 μL of BME to 40 mL of the Basal stripping buffer.
3. Incubate the membranes in this buffer at RT for 1 hour.
4. Rinse three times with dH_2O.
5. Wash three times with TBST, 5 minutes each wash, on a shaker.
6. Optional: Block for 30 minutes with 5% nonfat milk in TBST.

Notes:

- Prepare the basal stripping buffer and store at RT. Filtering is recommended to remove dust.
- Add BME just prior to use. Store BME stock at 4°C.
- Instead of BME, DTT can be used (10 mM final), although it's more expensive and has to be prepared from powder, aliquoted and stored at −20°C.
- The stripping time (step 3) can be shortened to 30 minutes, if the temperature is raised to 50°C.

Quick stripping protocol

Materials:

- Stripping buffer (2% SDS, 0.2 M NaOH).

Procedure:

1. Rinse the membranes in the Stripping buffer (5 seconds).

2. Incubate the membranes in the Stripping buffer at RT for 10 minutes, on a shaker.
3. Wash three times with TBST, 5 minutes each wash, on a shaker.
4. Optional: Block for 30 minutes with 5% nonfat milk in TBST.

Notes:

- This stripping protocol is best used for reprobing for a protein with a different molecular weight than the previous protein blotted for.
- If a phospho-blot is to be reprobed for total blot, pay attention to negative controls, since depending on the strength of the phospho-specific antibody, there could be some residue signal. In such cases, increase the stripping time from 10 to 15 minutes.

Coomassie Staining Protocol

Coomassie Blue Stain: 0.05% Coomassie Brilliant Blue R, 25% methanol, 10% acetic acid

Store at room temperature.

For 2 L of the stain:

1. Pour 500 mL of Methanol into a glass container
2. Add 1 g of Coomassie Brilliant Blue R
3. Stir (O/N).
4. Filter through Whatman filter paper.
5. Add 1300 mL of dH$_2$O
6. Add 200 mL of acetic acid (glacial).

Procedure:

1. Stain gels for 20min-1hr at room temperature (RT) on a shaker, in a covered container.
2. Destain with 20% methanol 10% acetic acid solution at RT, on a shaker, in a covered container until desired background is achieved.

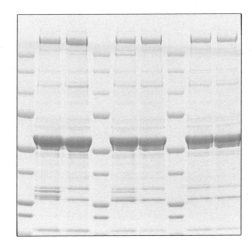

Lysis of Cells Using Native Conditions

The advantage of this method is that immunoprecipitation experiments can be performed after lysing the cells and protein quantification can be made using simple methods such as Bradford protein assay. An experimental procedure involving transfection and a GST pull-down will be described below, as an example for native lysis.

Materials:

- **1× Native Lysis Buffer (NLB):** 25 mM Hepes, pH 7.4, 1% Triton X-100 (or 0.2% NP-40 or 0.3% 3-[(3-cholamidopropyl)dimethylammonio]-1-propanesulfonate (CHAPS)), 120 mM NaCl, 0.27 M sucrose, 5 mM ethylenediaminetetraacetic acid (EDTA), 5 mM ethylene glycol-bis(β-aminoethyl ether)-N,N,N',N'-tetraacetic acid (EGTA), 50 mM NaF, 10 mM *b*-glycerophosphate, 5 mM sodium pyrophosphate, 2 mM benzamidine (fresh), 0.1% beta-mercaptoethanol (BME) (fresh) (or 1 mM dithiothreitol (DTT)), 1 mM phenylmethylsulfonyl fluoride (PMSF) (fresh), 2× Roche's complete protease inhibitor cocktail.
- 1× phosphate-buffered saline (PBS) (cold)
- Polyethylenimine (PEI) [polysciences, 24765-2 g, (MW 40,000)]
- Glutathione (GSH)-sepharose beads

Procedure:

1. Seed HEK293T cells into 15 cm dishes.
2. Next day, transfect with 20 μg plasmid + 55 μL PEI + 1 mL OptiMEM (incubate for 20 min @ RT).
3. 48 hours later, aspirate the medium. Wash with 10 mL PBS, on ice.
4. Lyze in 0.5−0.75 mL NLB per 10−15 cm dish.
5. Spin: 16,000 × g, 15 minutes, 4°C.
6. Take 100 μL out and add 20 μL of 5× Sodium dodecyl sulfate-polyacrylamide gel electrophoresis (SDS-PAGE) sample buffer. Freeze at −20°C.

7. Take 2, 4, 6 μL of the lysates and add 150 μL of 1× Bradford reagent. Measure OD595 to quantify the total protein concentration. Normalize the total protein concentration of the 120 μL lysates from step 6 using 1× SDS−PAGE sample buffer.
8. Add 20−30 μL GSH-sepharose beads per 2 mg of lysates.
9. Incubate at 4°C for 4 hours, on a wheel.
10. Wash: Four times with 1 mL of NLB.
11. Elute with 50 μL of 1× SDS−PAGE sample buffer or 1× SDS Lysis Buffer (SLB) (see Appendix I).
12. Samples can be frozen at this step at −20°C.
13. Heat at 70−90°C for 5−10 minutes (use 1000 rpm shaker for beads). Spin: $16,000 \times g$, 1 minutes, RT.
14. Load 15−20 μL lysate per lane or quantify the protein concentration and load 20 μg of lysate per lane.

Notes:

• If AMPK or other stress-response pathway is being studied, do not use cold PBS. Instead, pre-warm PBS to 37 °C to avoid stressing the cells.
• The strength of detergents that can be used for native lysis decreases from Triton X-100 to NP-40 to CHAPS. Triton X-100 would dissolve the nuclear envelope and most other membrane-bound organelles, while CHAPS would dissolve nuclear envelope minimally. For multi-protein complexes, Triton X-100 is usually too strong and either CHAPS or NP-40 should be employed. For single recombinant protein purification purposes, NP-40 and CHAPS are usually too mild and Triton X-100 should be employed.

APPENDIX *I*

Quick Denaturing Lysis Protocol

The advantage of this method is that it is rapid and uses denaturing conditions, in order to prevent loss of posttranslational modifications, such as phosphorylation, ubiquitination, etc. Moreover, it simplifies the procedure, potentially omitting the necessity to quantify protein concentration, for many applications.

	Number of Cells (Millions)	Lysis Buffer Volume (μL)	Total Protein (μg)
6-Well	0.9–1.2	350	300–400
12-Well	0.3–0.5	200	100–150
24-Well	0.1–0.15	100	40–60

Materials:

- **1 × SDS Lysis Buffer (SLB):** 63 mM Tris–HCl, pH 6.8, 2% sodium dodecyl sulfate (SDS), 1% beta-mercapto ethanol or 100 mM dithiothreitol (DTT), 0.01% bromophenol blue (BPB), 10% glycerol, 50 mM NaF, 2 mM Na_3VO_4
- 1 × phosphate-buffer saline (PBS) (cold)
- Sonicator or Benzonase

Procedure:

1. Aspirate the medium using a 1-mL pipette tip.
2. Rinse the cells with cold 1 × PBS, twice. Use 3 mL of PBS per 12-well plate, per wash.
3. Deliver 200 μL of 1 × SLB.
4. Shake at RT for 1 minute.
5. Collect into 1.5 mL tubes.
6. Sonicate at 50% power for 3–5 seconds or add Benzonase to 1:1000 dilution and incubate at RT for 3–5 minutes.
7. Samples can be frozen at this step at −20°C.
8. Heat at 70–90°C for 5–10 minutes. Spin: 16,000 × g, 1 minute, RT.
9. Load 15–20 μL lysate per lane or quantify the protein concentration and load 20 μg of lysate per lane.

Notes:

- If the cells are seeded into 12-well plates using a repeater pipette (i.e., consistent cell seeding and low variation between wells is ensured) and if the treatments are not expected to result in cell death or major protein synthesis inhibition (i.e., consistent total protein concentrations among samples), there may not be any need to quantify the protein levels and $15-20\ \mu L$ from a well of a 12-well plate (with 70%–80% confluent cells lyzed in $200\ \mu L$ SLB) will give about $20\ \mu g$ of total protein per lane.
- If protein levels are required to be quantified, use BCA protein assay.
- If Benzonase is not to be used, and protein quantification will not be made using BCA assay, one can supplement the SLB with 50 mM EDTA, in order to maximize the inhibition of phosphatases.
- If AMPK or other stress−response pathway is being studied, do not use cold PBS. Instead, prewarm PBS to 37°C to avoid stressing the cells.

Protein Tags

Flag tag (GATTACAAGGATGACGATGACAAG, DYKDDDDK)

HA tag (TACCCATACGACGTCCCAGACTACGCT, YPYDVPDYA)

Myc tag (GAACAAAAACTTATTTCTGAAGAAGATCTG, EQKL ISEEDL)

Pros:

- Small (8–9 amino acids)
- Can be placed N-terminally, C-terminally, or internally without loss of detectability by antibodies or major interference with protein function. (Note: Certain antibodies will detect only N- or C-terminal tags.)
- Due to small size, cloning is simplified by placing the sequence into the primers.
- Very specific and high affinity antibodies and antibody–bead conjugates are available.
- Commercially available antibody–bead conjugates help to avoid interference from the antibodies in the serum from the growth medium that may contaminate cell lysates.

Cons:

- Resin (antibody–bead conjugates or free antibody + protein A/G beads) is costly.
- Native elution from beads is costly (HA, myc, Flag, or 3xFlag peptide).
- Binding is weak in high stringency (high salt or high detergent concentration) or high ethylenediaminetetraacetic acid (EDTA) (for Flag) buffers (which is useful for protection of protein phosphorylation).
- Elution with sodium dodecyl sulfate polyacrylamide gel electrophoresis (SDS-PAGE) sample buffer results in elution of the antibody and interference with 50 and 25 kDa proteins during immunoblotting.

- Antibody—bead conjugates available do not always give clean pull-downs, depending on the stringency of the wash and commercial batch, due to a crosslinking step used to covalently conjugate the antibodies to the beads.

Glutathione S-Transferase (GST) tag

Pros:

- Glutathione-sepharose beads used to pull-down GST-tagged proteins are very specific, high affinity, and clean.
- Glutathione-sepharose beads are cheap and native elution from beads with free reduced glutathione is cheap and rapid (20 minutes at RT).
- No antibody heavy/light chains. Thus, no interference with 50 and 25 kDa proteins during immunoblotting, allowing elution with SDS—PAGE sample buffer.
- Increases solubility of the expressed proteins.

Cons:

- Large tag (26 kDa).
- Internal cloning is more likely to disrupt protein function due to large size of the tag.
- Special vectors with the GST tag open reading frame (ORF) are required.
- Potentially can interfere with the protein function, due to large size of the tag.
- Some versions of the GST tag may spontaneously dimerize.

Green Fluorescent Protein (GFP) tag

Pros:

- Very specific, high affinity, and clean llama antibody—bead conjugates are commercially-available (Chromotek).
- Elution from llama beads with SDS—PAGE sample buffer does not interfere with protein detection, since the llama antibodies are single chain and are ~15 kDa in size.
- The fluorescent tag can also be used for microscopy.
- The antibody—bead conjugates are very clean.

Cons:

- Large tag (27 kDa).
- Internal cloning is more likely to disrupt protein function due to large size of the tag.
- Special vectors with the GFP tag ORF are required.
- Potentially can interfere with the protein function, due to large size of the tag.
- Native elution from beads is very expensive (purified GFP has to be used).
- Some versions of the GFP tag may spontaneously dimerize.

His tag (CATCATCACCATCACCAT, HHHHHH)

Pros:

- Small tag (6 amino acid).
- Due to small size, cloning is simplified by placing the sequence into the primers.
- Native elution from beads is very cheap (imidazole or EDTA).
- No antibody heavy/light chains. Thus, no interference with 50 and 25 kDa proteins during immunoblotting, allowing elution with SDS−PAGE sample buffer.

Cons:

- Beads are usually not very clean and the Ni-NTA resin is usually not very specific.
- Antibodies for detection by immunoblotting are usually not very specific.
- The tag is not very specific—tandem six histidines are present endogenously in several proteins.

Covalent Crosslinking of Antibodies to Beads

(Adapted from Lamond Lab protocols—www.lamondlab.com)

Materials:

- BB—binding buffer (0.1 M sodium borate, pH 9.0)
- DMP (dimethyl pimelimidate, Sigma, D8388-250 mg). Prepare solutions fresh!
- Dissolve DMP in BB to 20 mM final concentration (5.2 mg in 1 mL).
- 50 mM glycine, pH 2.5.

Procedure:

1. Wash the protein G-sepharose beads three times, with 10 volumes of PBS, to remove ethanol.
2. Combine 200 μg of antibody with 100 μL of beads.
3. Incubate at 4°C for 4 hours or overnight.
4. Wash the beads four times, to remove unbound antibody.
5. Wash the beads two times with 10 volumes of BB.
6. Spin the beads. Remove the supernatant.
7. Resuspend in 10 volumes of fresh 20 mM DMP in BB.
8. Incubate at RT for 30 minutes on a wheel.
9. Repeat steps 6–8.
10. Spin the beads. Remove the supernatant.
11. Wash the beads two times, with 10 volumes of 50 mM glycine, pH2.5.
12. Wash the beads three times, with 10 volumes of PBS.
13. Resuspend the beads at a final density of 20%.
14. Store at 4°C for 1–2 months.

REFERENCES

Burnette, W.N., 1981. Anal. Biochem. 112, 195−203.

Fuhs, S.R., Meisenhelder, J., Aslanian, A., Ma, L., Zagorska, A., Stankova, M., et al., 2015. Monoclonal 1- and 3-phosphohistidine antibodies: new tools to study histidine phosphorylation. Cell 162 (1), 198−210.

Johnson, C., Crowther, S., Stafford, M.J., Campbell, D.G., Toth, R., MacKintosh, C., 2010. Bioinformatic and experimental survey of 14-3-3-binding sites. Biochem. J. 427 (1), 69−78. Available from: http://dx.doi.org/10.1042/BJ20091834.

Mackintosh, C., 2004. Dynamic interactions between 14-3-3 proteins and phosphoproteins regulate diverse cellular processes. Biochem. J. 381 (Pt 2), 329−342.

Moorhead, G., Douglas, P., Morrice, N., Scarabel, M., Aitken, A., MacKintosh, C., 1996. Phosphorylated nitrate reductase from spinach leaves is inhibited by 14-3-3 proteins and activated by fusicoccin. Curr. Biol. 6, 1104−1113.

Southern, E.M., 1975. J. Mol. Biol. 98, 503−517.

Towbin, H., Staehelin, T., Gordon, J., 1979. Proc. Natl. Acad. Sci. USA 76, 4350−4354.

⟨www.kinase.com⟩

⟨www.LamondLab.com⟩

FURTHER READING

Alwine, J.C., Kemp, D.J., Stark, G.R., 1977. Proc. Natl. Acad. Sci. USA 74, 5350−5354.

Reiser, J., Stark, G., 1979. Proc. Natl. Acad. Sci. USA 76, 3116−3120.